Study Guide
for Moore, McCabe, and Craig's

Introduction to the Practice of Statistics

Sixth Edition

Michael A. Fligner
Ohio State University

W. H. Freeman and Company
New York

ISBN-13: 978-1-4292-1473-5
ISBN-10: 1-4292-1473-2

Printed in the United States of America

First printing

W. H. Freeman and Company
41 Madison Avenue
New York, NY 10010
Houndmills, Basingstoke
RG21 6XS, England

CONTENTS

Preface

Chapter 1: Looking at Data – Distributions 1

Chapter 2: Looking at Data – Relationships 31

Chapter 3: Producing Data 64

Chapter 4: Probability: The Study of Randomness 83

Chapter 5: Sampling Distributions 110

Chapter 6: Introduction to Inference 127

Chapter 7: Inference for Distributions 145

Chapter 8: Inference for Proportions 163

Chapter 9: Analysis of Two-Way Tables 176

Chapter 10: Inference for Regression 190

Chapter 11: Multiple Regression 202

Chapter 12: One-Way Analysis of Variance 210

Chapter 13: Two-Way Analysis of Variance 224

Chapter 14: Logistic Regression 233

Chapter 15: Nonparametric Tests 241

Chapter 16: Bootstrap Methods and Permutation Tests 250

Chapter 17: Statistics for Quality: Control and Capability 261

PREFACE TO THE STUDY GUIDE

This study guide has been written to help you learn and review the material in the text. The structure of this study guide is as follows: We first provide an overview of each section, which reviews the key concepts of that section. After the overview, there are *guided* solutions to selected odd-numbered problems in that section, along with the key concepts from the section required for solving each problem. The guided solutions provide hints for setting up and thinking about the exercise, which should help you improve your problem-solving skills. Once you have worked through the guided solution, you can look up the complete solution provided later in the study guide to check your work.

What is the best way to use this study guide? Part of learning involves doing homework problems. In doing a problem, it is best for you to first try to solve the problem on your own. If you are having difficulty, then try to solve the problem using the hints and ideas in the guided solution. If you are still having difficulty, then you can read through the complete solution. Generally, try to use the complete solution as a way to check your work. Be careful not to confuse reading the complete solutions with doing the problems themselves. This is the same mistake as reading a book about swimming and then believing you are prepared to jump into the deep end of the pool. If you simply read our complete solutions and convince yourself that you could have worked the problem on your own, you may be misleading yourself and could have trouble on exams. When you are having difficulty with a particular type of problem, find similar problems to work in the exercises in the text. Often problems adjacent to each other in the text use the same ideas.

In the overviews we try to summarize the material that is most important, which should help you review material when you are preparing for a test. If any of the terms in the overview are unfamiliar, you probably need to go back to the text and reread the appropriate section. If you are having difficulty with the material, do not neglect to see your instructor for help. Face-to-face communication with your instructor is the best way to clear up any confusion.

CHAPTER 1

LOOKING AT DATA—DISTRIBUTIONS

SECTION 1.1

OVERVIEW

Section 1.1 introduces several methods for exploring data. These methods should only be applied after clearly understanding the background of the data collected. The choice of method depends to some extent upon the type of variable being measured. The two types of variables described in this section are

- **Categorical variables:** variables that record to what group or category an individual belongs. Hair color and gender are examples of categorical variables. Although we might count the number of people in the group with brown hair, we wouldn't think about computing an average hair color for the group, even if numbers were used to represent the hair color categories.

- **Quantitative variables:** variables that have numerical values and with which it makes sense to do arithmetic. Height, weight, and GPA are examples of quantitative variables. It makes sense to talk about the average height or GPA of a group of people.

To summarize the **distribution** of a variable, for categorical variables use **bar graphs** or **pie charts,** while for numerical data use **histograms** or **stemplots.** Also, when numerical data are collected over time, in addition to a histogram or stemplot, a **timeplot** can be used to look for interesting features of the data. When examining the data through graphs we should be on the alert for

- unusual values that do not follow the pattern of the rest of the data.
- some sense of a central or typical value of the data.
- some sense of how spread out or variable the data are.
- some sense of the shape of the overall pattern.

In addition, when drawing a timeplot be on the lookout for **trends** occurring over time. Although many of the graphs and plots may be drawn by computer, it is still up to you to recognize and interpret the important features of the plots and the information they contain.

GUIDED SOLUTIONS

Exercise 1.3

KEY CONCEPTS: Individuals and type of variables

When identifying the individuals or cases described, you need to include sufficient detail so that it is clear which individuals are contained in the data set. Describe the individuals or cases below.

Recall that the variables are the characteristics of the individuals. How many variables are there? Once the variables are identified, you need to determine if they are categorical (the variable puts individuals or objects into one of several groups) or quantitative (the variable takes meaningful numerical values for which arithmetic operations make sense). Now list the variables recorded and classify each as categorical or quantitative.

Name of variable	Type of variable

Exercise 1.19

KEY CONCEPTS: Drawing bar graphs and Pareto charts

(a) Complete the bar graph below. The first and last bars have been completed for you. The software package has arranged the categories in alphabetical order as the default.

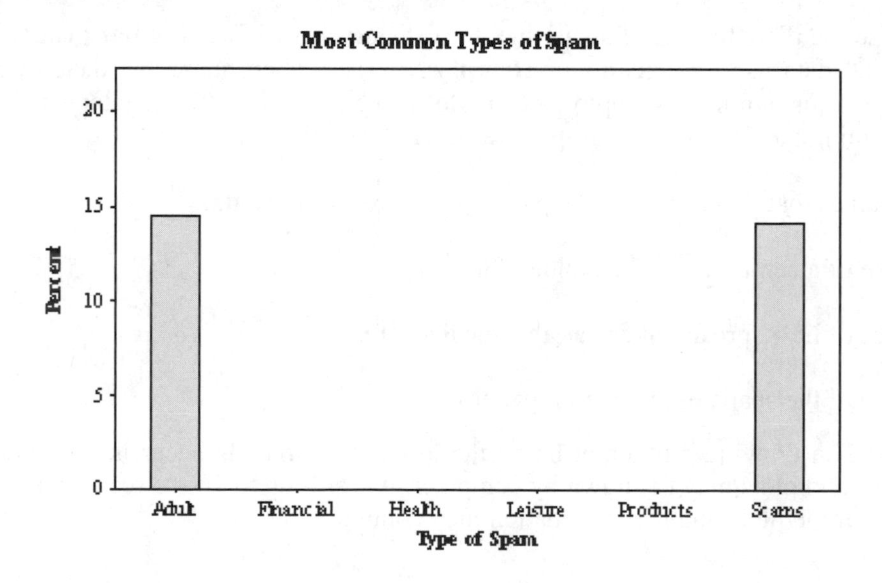

(b) Many software packages will order the categories by the bar heights for you. This can be done in increasing or decreasing order and makes it easier to compare categories, particularly those that have similar percentages. The first two bars have been drawn for you. Complete the Pareto chart below.

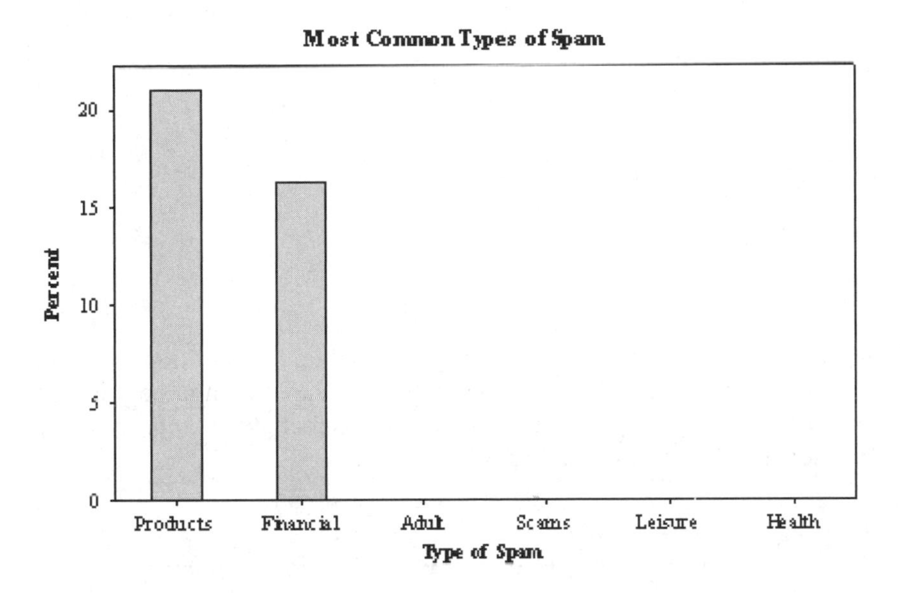

Exercise 1.27

KEY CONCEPTS: Interpreting a histogram

How would you describe this distribution? Which portion of the histogram do you think corresponds to the state schools? How about the more exclusive private schools? In general, how many groups of schools are there and what are the most important aspects of the distribution?

Exercise 1.40

KEY CONCEPTS: Drawing histograms and stem-and-leaf plots and interpreting their shapes

Density measurements:

```
5.50   5.61   4.88   5.07   5.26   5.55   5.36   5.29   5.58   5.65
5.57   5.53   5.62   5.29   5.44   5.34   5.79   5.10   5.27   5.39
5.42   5.47   5.63   5.34   5.46   5.30   5.75   5.68   5.85
```

How to draw a histogram:

i) When drawing a histogram, choose class intervals that divide the data into classes of equal length. For this data set, the smallest density measurement is 4.88 and the largest is 5.85, so the class intervals need to cover this entire range. A simple set of class intervals would be 4.70 to <4.90 (4.70 is included in the interval but not 4.90), 4.90 to <5.10, 5.10 to <5.30, 5.30 to <5.50 , 5.50 to <5.70, and 5.70 to <5.90. Other sets of intervals are possible, although these have the advantage of using fairly simple numbers as endpoints. Computer software will automatically choose class intervals for you, but most software also provides the option of specifying your own intervals.

ii) Count the number of data values in each class interval. Using the class intervals above, complete the frequency table below. Remember, when counting the number in each interval be sure to include data values equal to the lower endpoint but not the upper endpoint.

Interval Count Percent

iii) Draw the histogram, which is a picture of the frequency table. In addition to drawing the bars, this also requires labeling the axes. Using the frequency table you have computed, complete the histogram given. The scale for the *x* axis has been included. Make sure to include an appropriate label for the *x* and *y* axes, as well as numbers for the scale on the *y* axis.

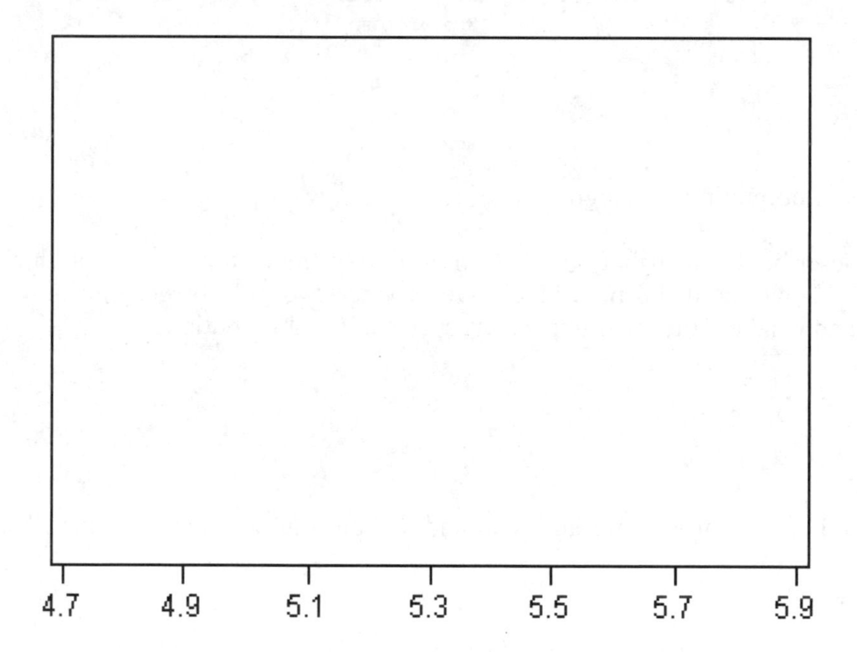

How to draw a stem-and-leaf plot—single stems and splitting stems:

i) It is easiest, although not necessary, to first order the data. If the data have been ordered, the leaves on the stems will be in increasing order. The density measurements have been ordered for you below.

```
4.88   5.07   5.10   5.26   5.27   5.29   5.29   5.30   5.34   5.34
5.36   5.39   5.42   5.44   5.46   5.47   5.50   5.53   5.55   5.57
5.58   5.61   5.62   5.63   5.65   5.68   5.75   5.79   5.85
```

ii) Using the stems below, complete the stem-and-leaf plot.

```
4.8
4.9
5.0
5.1
5.2
5.3
5.4
5.5
5.6
5.7
5.8
```

What are the similarities between this stem-and-leaf plot and the histogram you drew? Think about the relationship between the stems and the class intervals in this example, which helps to explain why a stem-and-leaf plot looks like a histogram laid on its side. What class intervals do the stems correspond to? What is one important difference between the histogram and the stem-and-leaf plot?

To finish up the exercise, think about the important features that describe a distribution. Does the distribution of the density measurements have a single peak? Does it appear to be symmetric, or is it skewed to the right (tail with larger values is longer), or to the left? Are there any outliers that fall outside the overall pattern of the data? What is your estimate of the density of the earth based on these measurements?

Exercise 1.46

KEY CONCEPTS: Drawing and interpreting a time plot

Complete the time plot on the graph below. The winning times for 1972, 1973, and 1974 are included in the plot to get you started.

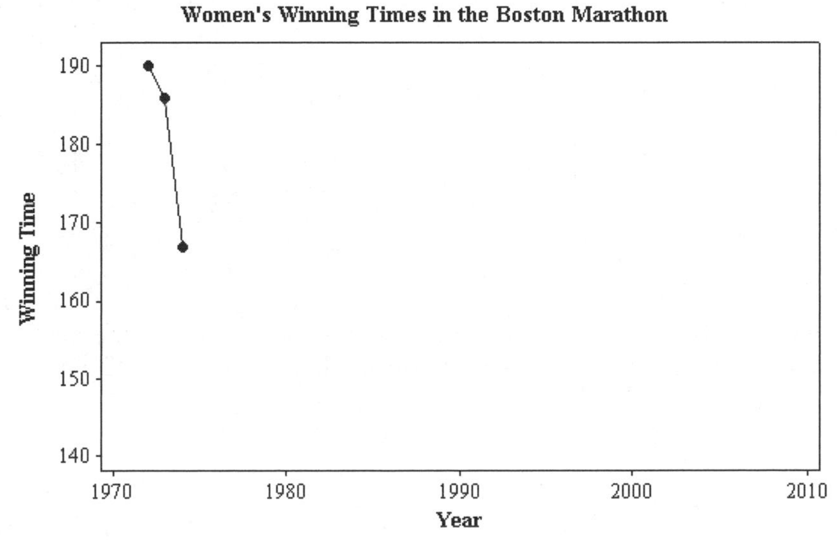

What is the general pattern in the time plot? Have times stopped improving in recent years?

Complete Solutions

Exercise 1.3

The individuals or cases in this exercise are the different apartments.

There are 5 variables which are "monthly rent," "free cable," "pets allowed," "number of bedrooms," and "distance from campus." The variable types are "monthly rent" (quantitative), "free cable" (categorical with categories yes or no), "pets allowed" (categorical with categories yes or no), number of bedrooms (probably can be considered as categorical), and "distance from campus" (quantitative). Although the variable "number of bedrooms" is recorded as a number, this variable divides apartments into only a few categories. We could compute the average number of bedrooms per apartment, treating it as a quantitative variable, but we might also be interested the percent of 2, 3, or 4 bedroom apartments, treating it as a categorical variable.

.

Exercise 1.19

(a) The complete bar graph is given below. Note that the comparison of similar (in terms of percents) categories such as "Adult"and "Scams" is difficult with the bars far apart.

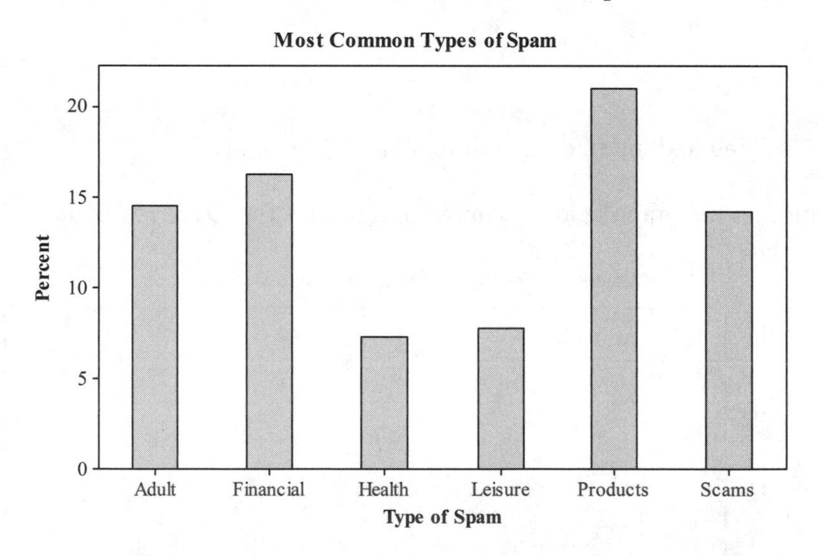

(b) The Pareto chart with bars in order from tallest to shortest is given on the next page. Note that the comparison of similar (in terms of percents) categories such as "Adult"and "Scams" is much easier with the bars adjacent.

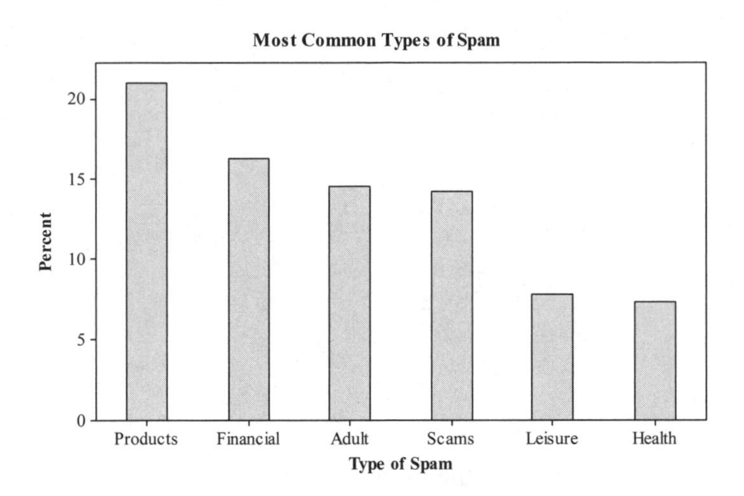

Exercise 1.27

There are three groups of schools. The state schools such as the University of Massachusetts have the lower tuitions and form the group with tuitions of $10,000 and below. The private schools seem to be divided into two groups. There are 23 schools in the range $12,000 to $22,000 and they include less expensive private schools such as Northeastern University. At the high end of the distribution (over $28,000) are some of the most expensive private colleges, including, for example, Harvard University. This distribution provides an example of a trimodal (three modes) distribution that has been created by including three distinct groups of schools in the distribution.

Exercise 1.40

Interval	Count	Percent
4.70-<4.90	1	3.45%
4.90-<5.10	1	3.45%
5.10-<5.30	5	17.24%
5.30-<5.50	9	31.03%
5.50-<5.70	10	34.48%
5.70-<5.90	3	10.34%
	29	99.99%

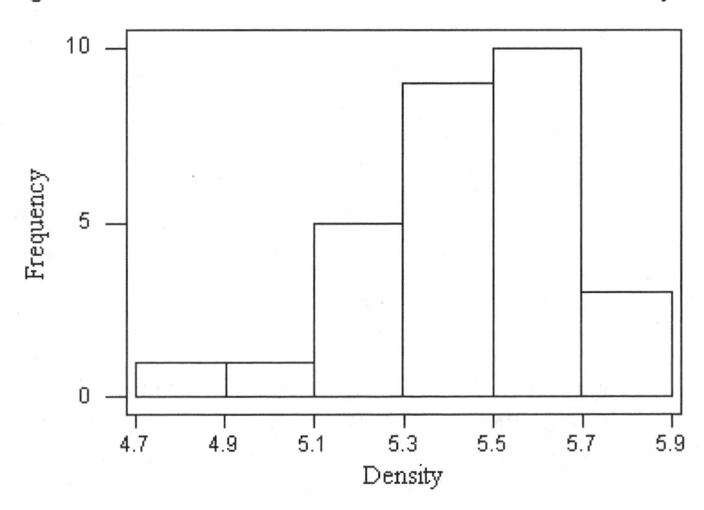

The stemplot is given below.

```
4.8 | 8
4.9 |
5.0 | 7
5.1 | 0
5.2 | 6799
5.3 | 04469
5.4 | 2467
5.5 | 03578
5.6 | 12358
5.7 | 59
5.8 | 5
```

In general, for this number of observations, the preference for a histogram over a stem-and-leaf plot is a personal preference. For smaller data sets, the stem-and-leaf plot is usually preferred, while for larger data sets most people prefer the histogram. When a data set starts to get large the stemplot can begin to look a little cluttered. However, in this particular problem, the general shape of the distribution is quite apparent from either plot.

Further comments on the general shape: the stemplot and histogram give a distribution that appears fairly symmetric, although the histogram looks a little more skewed. There is one possible outlier of 4.88, but this is not that far from the bulk of the data. Because of the choice of intervals in the histogram, the 4.88 does not appear as far removed from the bulk of the data in the histogram as it does in the stemplot. An estimate of the density would be the center of the distribution, probably somewhere between 5.4 and 5.5. In the next section, more formal numerical methods for estimating the density from the distribution will be described.

Exercise 1.46

The complete time plot is given below.

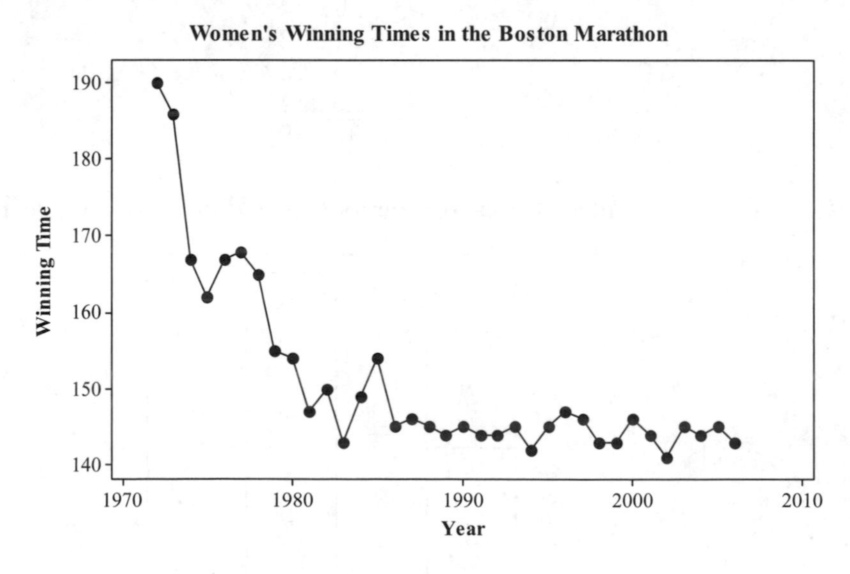

There is a fairly steady downward trend in the winning times until around 1982. After this point the improvement seems to have stopped, with the winning times varying around 145 minutes. The variability in the winning times from year to year has also gone down.

SECTION 1.2

OVERVIEW

Although graphs give an overall sense of the data, numerical summaries of features of the data make more precise the notions of center and spread.

Two important measures of center are the **mean** and the **median.** If there are n observations, $x_1, x_2, \cdots, x_n$, then the mean is

$$\bar{x} = \frac{x_1 + x_2 + \cdots + x_n}{n} = \frac{1}{n}\sum x_i$$

where $\sum$ means "add up all these numbers." Thus, the mean is just the total of all the observations divided by the number of observations.

While the median can be expressed by a formula, it is simpler to describe the rules for finding it.

How to find the median:

1. List all the observations from smallest to largest.

2. If the number of observations is odd, then the median is the middle observation. Count from the bottom of the list of ordered values up to the $(n + 1)/2$ largest observation. This observation is the median.

3. If the number of observations is even, then the median is the average of the two center observations.

The most important measures of spread are the **quartiles,** the **standard deviation**, and the **variance.** For measures of spread, the quartiles are appropriate when the median is used as a measure of center. In general, the median and quartiles are more appropriate when outliers are present or when the data are skewed. In addition, the **five-number summary,** which reports the largest and smallest values of the data, the quartiles, and the median, provides a compact description of the data that can be represented graphically by a **boxplot.** Computationally, the first quartile, Q_1, is the median of the lower half of the list of ordered observations, and the third quartile, Q_3, is the median of the upper half of the list of ordered values.

If you use the mean as a measure of center, then the standard deviation and variance are the appropriate measures of spread. Remember that means and variances can be strongly affected by outliers and are harder to interpret for skewed data.

If we have n observations, $x_1, x_2, \cdots, x_n$, with mean $\bar{x}$, then the variance s^2 can be found using the formula

$$s^2 = \frac{(x_1 - \bar{x})^2 + (x_2 - \bar{x})^2 + \cdots + (x_n - \bar{x})^2}{n-1} = \frac{1}{n-1}\sum (x_i - \bar{x})^2$$

The standard deviation is the square root of the variance, i.e., $s = \sqrt{s^2}$, and is a measure of spread in the same units as the original data. If the observations are in feet, then the standard deviation is in feet as well.

GUIDED SOLUTIONS

Exercise 1.63

KEY CONCEPTS: Measures of center, five-number summary

(a) The five-number summary consists of the median, the minimum and the maximum, and the first and third quartiles. To compute these quantities, it is simplest to first order the data. The ordered average property damages ($millions) are given below. There are 51 observations as the data includes all 50 states and Puerto Rico. What is the location of the median? Locate and underline the median in the list below.

```
 0.00    0.05    0.09    0.10    0.24    0.26    0.27    0.34    0.53
 0.66    1.49    1.78    2.14    2.26    2.27    2.33    2.37    2.94
 3.47    3.57    3.68    4.42    4.62    5.52    7.42   10.64   14.69
14.90   15.73   17.11   17.19   23.47   24.84   27.75   29.88   30.26
31.33   37.32   40.96   43.62   44.36   49.28   49.51   51.68   51.88
53.13   62.94   68.93   81.94   84.84   88.60
```

Now, remember that the first quartile is the median of the 25 observations below the median, and the third quartile is the median of the observations above the median. Underline the quartiles and fill in the table below.

Minimum
Q_1
M
Q_3
Maximum

Explain why you can see from these numbers that the distribution is right-skewed.

(b) The histogram of the data is given below.

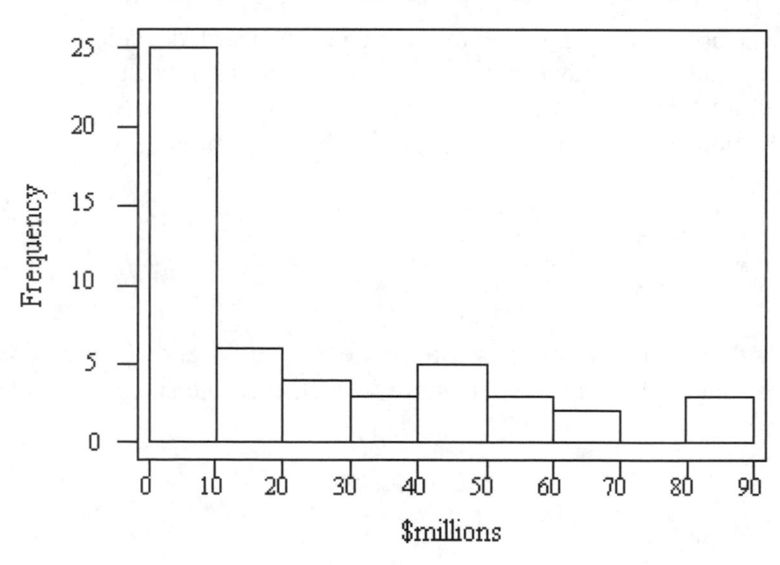

The histogram suggests there are a few large outliers, namely the states above 80. To determine if the $1.5 \times IQR$ criterion flags the largest states, first compute the IQR. It is $Q_3 - Q_1$.

$IQR =$ $1.5 \times IQR =$

Then add $1.5 \times IQR$ to the third quartile and see if the damage from the largest state exceeds this value.

$$Q_3 + (1.5 \times IQR) =$$

You should find that there are no suspected outliers in the data according to this criterion.

(c) To find the mean, find the sum of the 51 observations, then divide by the number of observations.

Sum

Mean

Explain why the mean and median differ so greatly for this distribution.

Exercise 1.67

KEY CONCEPTS: Measures of center

When there are several observations at a single value, the key is to remember that the mean is the total of all the observations divided by the number of observations. When computing the total, remember to include a salary as many times as it appears. The same is true when ordering the observations to find the median: remember to include a salary as many times as it appears.

Exercise 1.69

KEY CONCEPTS: Measures of center, resistant measures

The change of extremes affects the mean, but not the median. To compute the new mean you can figure out the new total (you don't need to add up all the numbers again—just think about how much it has gone up) and divide by the number of observations. Or else you can think about dividing up the salary increases by the number of observations and adding this value to the old mean.

Exercise 1.89

KEY CONCEPTS: Standard deviation

There are two points to remember in finding the answer. The first is that numbers "farther apart" from each other tend to have higher variability than numbers closer together. The other is that repeats are allowed. There are several choices for the answer to (a) but only one for (b).

Exercise 1.94

KEY CONCEPTS: Mean and standard deviation, linear transformations

(a) Given the distribution of measurements (see the histogram in Exercise 1.40), $\bar{x}$ and s are good measures to describe this distribution. Find their values and give an estimate of the density of the earth based on these measures. Remember, if you are doing the calculation by hand, first find $\bar{x}$ and then calculate s in steps.

(b) This is an exercise in recognizing when a new measurement can be expressed in terms of an old measurement by the equation $x_{new} = a + bx$ and then seeing the effect of this linear transformation on measures of center and spread. This form of transforming an old measurement to a new measurement is called a linear transformation.

Express Cavendish's first observation of 5.50 g/cm^3 in pounds per cubic foot.

_____________ pounds per cubic foot.

What values of a and b are being used for x_{new}? (Hint: In this exercise, the value of a is equal to zero.)

Now, using the rules given for the effect of a linear transformation on measures of center and spread, give the values of $\bar{x}$ and s in pounds per cubic foot.

$\bar{x}$ = __________ pounds per cubic foot

s = __________ pounds per cubic foot

COMPLETE SOLUTIONS

Exercise 1.63

(a) The ordered observations are reproduced below. Since there are 51 observations, the $(n + 1)/2$ largest observation is the $(51 + 2)/2 = 26^{th}$ largest observation. The median is given in bold and underlined. The first quartile is the median of the 25 observations below the median and is the $(n + 1)/2 = 13^{th}$ largest observation in this group. It is given in italics and underlined. Similarly, the third quartile is the 13th largest observation among the 25 observations above the median, and it is also given in italics and underlined.

0.00	0.05	0.09	0.10	0.24	0.26	0.27	0.34	0.53
0.66	1.49	1.78	*2.14*	2.26	2.27	2.33	2.37	2.94
3.47	3.57	3.68	4.42	4.62	5.52	7.42	**10.64**	14.69
14.90	15.73	17.11	17.19	23.47	24.84	27.75	29.88	30.26
31.33	37.32	*40.96*	43.62	44.36	49.28	49.51	51.68	51.88
53.13	62.94	68.93	81.94	84.84	88.60			

The five-number summary is given on the next page.

Minimum	0.00
Q_1	2.14
M	10.64
Q_3	40.96
Maximum	88.60

You can see from the five-number summary that the smallest quarter of the observations are between 0 and 2 and the smaller half of the observations are between 0 and 10. The top half of the distribution is much more spread out, suggesting very strong skewness to the right.

(b) The IQR is 40.96 – 2.14 = 38.82, and 1.5 × IQR = 58.23. If we add 58.23 to the third quartile, we get 40.96 + 58.23 = 99.19. Since this exceeds the maximum, no states are flagged as suspected outliers.

(c) The sum of the 51 observations is 1119.60, the mean is $\bar{x}$ = 1119.60/51 = 21.95, and the median is M = 10.64. The distribution is strongly right-skewed, and the mean exceeds the median because of this.

Exercise 1.67

The number of observations (individuals) is 5 + 2 + 1 = 8 and the total of the salaries is

$$(5 \times \$35{,}000) + (2 \times \$80{,}000) + (1 \times \$320{,}000) = \$655{,}000$$

The mean is $655,000/8 = $81,875. Everyone earns less than the mean except for the owner. Since there are eight observations, the median is the average of the fourth and fifth smallest observations, which is $35,000. To see this, the eight ordered observations are

$35,000, $35,000, $35,000, $35,000, $35,000, $80,000, $80,000, $320,000

Exercise 1.69

The owner has an increase in salary from $320,000 to $455,000, or an increase of $135,000. The total is increased by this amount, from $655,000 to $790,000 and the new mean is $790,000/8 = $98,750. Another way of thinking about it is that the $135,000 increase averaged among the 8 people is $16,875, so the mean must go up $16,875 to $81,875 + $16,875 = $98,750.

The fourth and fifth smallest observations are still the same, so the median is unaffected.

Exercise 1.89

(a) The standard deviation is always greater than or equal to zero. The only way it can equal zero is if all the numbers in the data set are the same. Since repeats are allowed, just choose all four numbers the same to make the standard deviation equal to zero. Examples are 1, 1, 1, 1 or 2, 2, 2, 2.

(b) To make the standard deviation large, numbers at the extremes should be selected. So you want to put the four numbers at 0 or 20. The correct answer is 0, 0, 20, 20. You might have thought 0, 0, 0, 20 or 0, 20, 20, 20 would be just as good, but a computation of the standard deviation of these choices shows that two at either end is the best choice.

(c) There are many choices for (a) but only one for (b).

Exercise 1.94

(a) In practice, you will be using software or your calculator to obtain the mean and standard deviation from keyed-in data. However, we illustrate the step-by-step calculations to help you understand how the standard deviation works. Be careful not to round off the numbers until the last step, as this can sometimes introduce fairly large errors when computing s.

Observation x_i	Difference $x_i - \bar{x}$	Difference squared $(x_i - \bar{x})^2$
5.50	0.052100	0.002714
5.61	0.162100	0.026277
4.88	-0.567900	0.322510
5.07	-0.377900	0.142808
5.26	-0.187900	0.035306
5.55	0.102100	0.010424
5.36	-0.087900	0.007726
5.29	-0.157900	0.024932
5.58	0.132100	0.017450
5.65	0.202100	0.040845
5.57	0.122100	0.014908
5.53	0.082100	0.006740
5.62	0.172100	0.029618
5.29	-0.157900	0.024932
5.44	-0.007900	0.000062
5.34	-0.107900	0.011642
5.79	0.342100	0.117033
5.10	-0.347900	0.121034
5.27	-0.177900	0.031648
5.39	-0.057900	0.003352
5.42	-0.027900	0.000778
5.47	0.022100	0.000488
5.63	0.182100	0.033161
5.34	-0.107900	0.011642
5.46	0.012100	0.000146
5.30	-0.147900	0.021874
5.75	0.302100	0.091265
5.68	0.232100	0.053870
5.85	0.402100	0.161684
Column sums = 157.99		1.366869

The mean is 157.99/29 = 5.4479. This has been subtracted from each observation to give the column of deviations or differences, $x_i - \bar{x}$. This column is squared to give the squared deviations $(x_i - \bar{x})^2$, in the third column. The variance is the sum of these squared deviations divided by 1 less than the number of observations:

$$s^2 = \frac{1.366869}{29-1} = 0.0488168 \text{ and } s = \sqrt{0.0488168} = 0.22095$$

Use the mean of 5.4479 as the estimate of the density of the earth based on these measurements, since the distribution is approximately symmetric and without outliers.

(b) Since the density of water is 1 gram per cubic centimeter or 62.43 pounds per cubic foot, Cavendish's first result of 5.50 g/cm^3 expressed in pounds per cubic foot is $5.50 \times 62.43 = 343.365$ pounds per cubic foot. The transformation we are using has a value of $a = 0$ and $b = 62.43$. Simply put, we are multiplying each observation by 62.43 to convert from g/cm^3 to pounds per cubic foot.

Using the rules given for the effect of a linear transformation on measures of center and spread, multiplying each observation by a positive number b multiplies both measures of center and measures of spread by b. Using your answers in (a),

$$\bar{x} = 5.4479 \times 62.43 = 340.11 \text{ pounds per cubic foot}$$

$$s = 0.22095 \times 62.43 = 13.79 \text{ pounds per cubic foot}$$

SECTION 1.3

OVERVIEW

This section considers the use of mathematical models to describe the overall pattern of a distribution. A mathematical model is an idealized description of this overall pattern, often represented by a smooth curve. The name given to a mathematical model that summarizes the shape of a histogram is a **density curve.** The density curve is a kind of idealized histogram. The total area under a density curve is 1 and the area between two numbers represents the proportion of the data that lie between these two numbers. Like a histogram, it can be described by measures of center, such as the **median** (a point such that half the area under the density curve is to the left of the point) and the **mean** μ (the balance point of the density curve if the curve were made of solid material), and measures of spread, such as the **quartiles** and the standard deviation σ.

One of the most commonly used density curves in statistics is the **normal curve.** The distributions they describe are called **normal distributions.** Normal curves are symmetric and bell-shaped. The peak of the curve is located above the mean and median, which are equal since the density curve is symmetric. The standard deviation is the distance from the mean to the change-of-curvature points on either side. It measures how concentrated the area is around the peak of the curve. Normal curves follow the **68-95-99.7 rule,** i.e., 68% of the area under a normal curve lies within one standard deviation of the mean (illustrated in the figure below), 95% within two standard deviations of the mean, and 99.7% within three standard deviations of the mean.

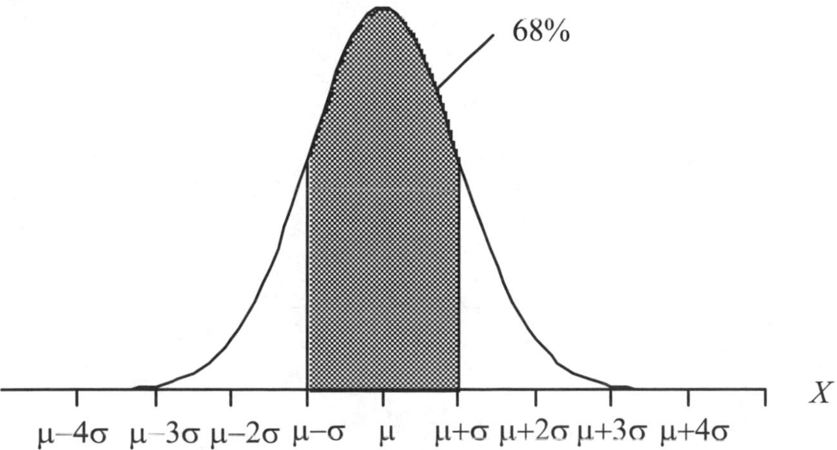

Areas under any normal curve can be found easily if quantities are first **standardized** by subtracting the mean from each value and then dividing the result by the standard deviation. This standardized value is sometimes called the ***z*-score.** If data whose distribution can be described by a normal curve are standardized (all values replaced by their z-scores), the distribution of these standardized values is called the **standard normal distribution** and they are described by the **standard normal curve.** Areas under standard normal curves are easily computed by using a standard normal table such as that found in Table A in the front inside cover of the text and in the back of the book.

If we know that the distribution of data is described by a normal curve, we can make statements about what values are likely and unlikely without actually observing the individual values of the data. Although one can examine a histogram or stem-and-leaf plot to see if it is bell-shaped, the preferred method for determining if the distribution of data is described by a normal curve is a **normal quantile plot.** These are easily made using modern statistical computer software. If the distribution of data is described by a normal curve, the normal quantile plot should look like a straight line.

In general, density curves are useful for describing distributions. Many statistical procedures are based on assumptions about the nature of the density curve that describes the distribution of a set of data. **Density estimation** refers to techniques for finding a density curve that describes a given set of data.

GUIDED SOLUTIONS

Exercise 1.109

KEY CONCEPTS: Density curves and area under a density curve

(a) In this case, the density curve has unknown height h between 0 and 4, and height 0 elsewhere. Thus, the density curve forms a rectangle with a base of length 4 and a height equal to h. Recall that the area of any rectangle is the product of the length of the base of the rectangle and the height. The area of this density curve is therefore (fill in the blanks)

area = length of base × height = ___________ × ___________

The total area under a density curve must be 1. What must h be so that this is the case?
Draw a graph of this density curve in the space provided.

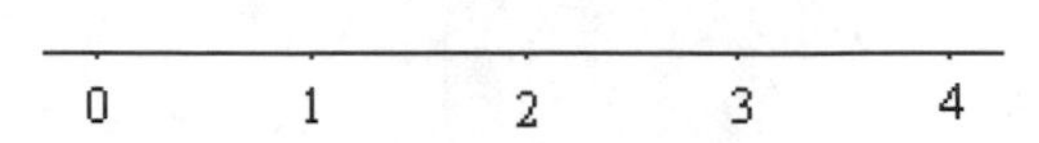

(b) The area of interest is the shaded region in the figure on the next page.

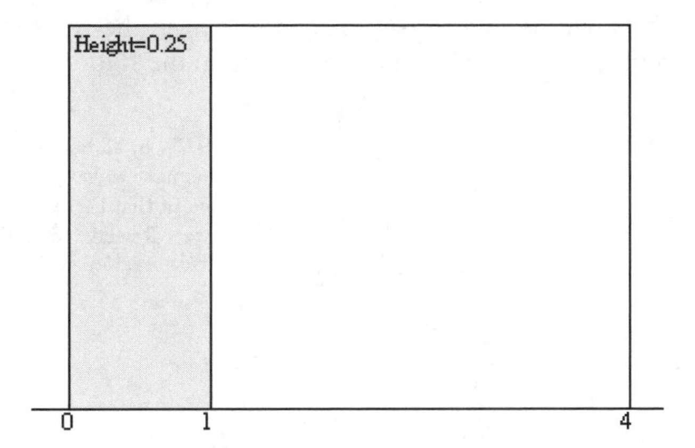

Compute the area of the shaded region by filling in the blanks below.

area = length of base × height = ________ × ________ = __________

(c) Try answering this part on your own, using the same reasoning as in (b). First, shade in the area of interest in the figure below.

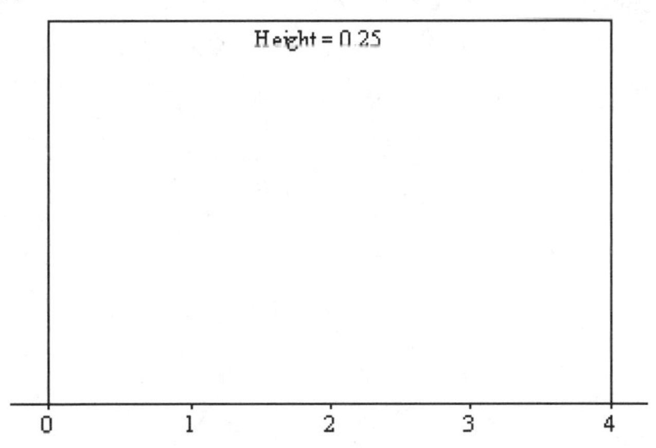

Now compute the area of your shaded region as in (b).

Exercise 1.115

KEY CONCEPTS: The 68-95-99.7 rule for normal curves

Recall that the 68-95-99.7 rule says that for the normal distribution, approximately 68% of the observations fall between the mean minus one standard deviation and the mean plus one standard deviation, 95% of the observations fall between the mean minus two standard deviations and the mean plus two standard deviations, and approximately 99.7% of the observations fall between the mean minus three standard deviations and the mean plus three standard deviations. Also recall that the area under a density curve between two numbers corresponds to the proportion of the data that lies between these two numbers.

In this problem the mean is 336 days and the standard deviation is 3 days. From the 68-95-99.7 rule we have, for example, that 68% of the lengths of all horse pregnancies lie between $336 - (1 \times 3) = 333$ and $336 + (1 \times 3) = 339$ days.

(a) The 68-95-99.7 rule says that approximately 99.7% of the observations fall between the mean minus three standard deviations and the mean plus three standard deviations. The mean minus three standard deviations is 327. This is the lower bound for the shaded region in the figure below. What is the upper bound for the shaded region below? Fill in the space provided in the figure.

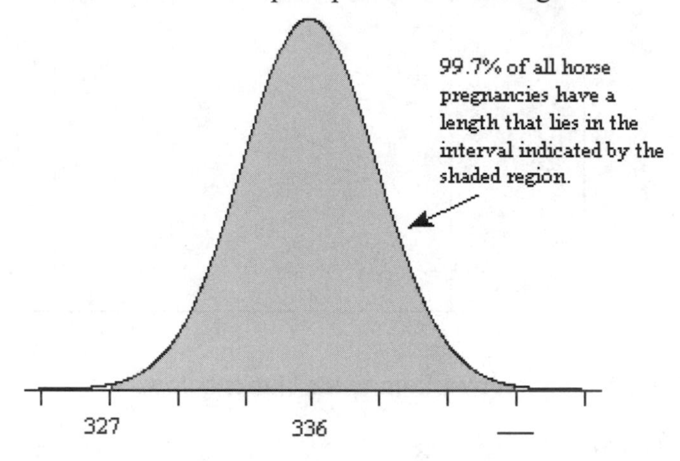

(b) We indicated above that the middle 68% of all horse pregnancies have lengths between 333 and 339 days. What percent are *either* less than 333 or longer than 339? What percent must therefore be longer than 339? (Recall that the density curve is symmetric.)

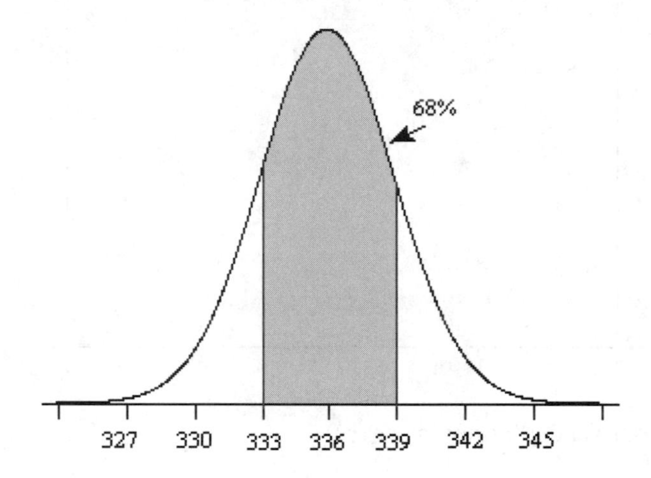

Exercise 1.121

KEY CONCEPTS: Computing relative frequencies for a standard normal distribution

Recall that the proportion of observations from a standard normal distribution that are less than a given value z is equal to the area under the standard normal curve to the left of z. Table A gives these areas. This is illustrated in the figure on the next page.

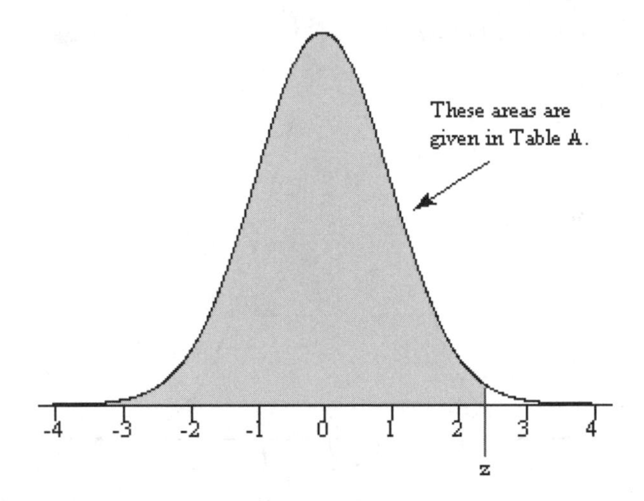

In answering questions concerning the proportion of observations from a standard normal distribution that satisfy some relation, we find it helpful to first draw a picture of the area under a normal curve corresponding to the relation. We then try to visualize this area as a combination of areas of the form in the figure above, since such areas can be found in Table A. The entries in Table A are then combined to give the area corresponding to the relation of interest.

This approach is illustrated in the solutions that follow.

(a) To get you started, we will work through a complete solution. A picture of the desired area is given below.

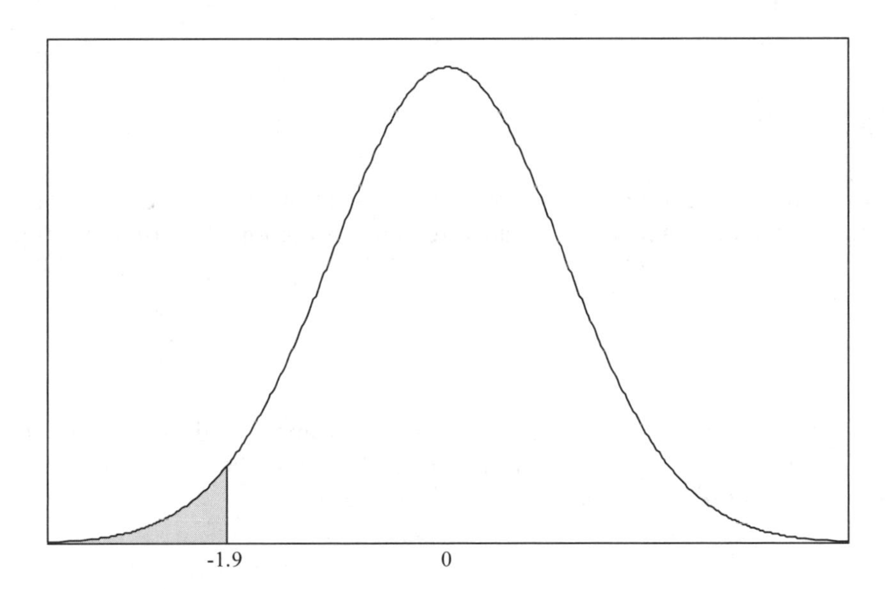

This is exactly the type of area that is given in Table A. We simply find the row labeled –1.9 along the left margin of the table, locate the column labeled .00 across the top of the table, and read the entry in the intersection of this row and column. This entry is 0.0287. This is the proportion of observations from a standard normal distribution that satisfies $z < -1.90$.

(b) Shade the desired area in the figure below.

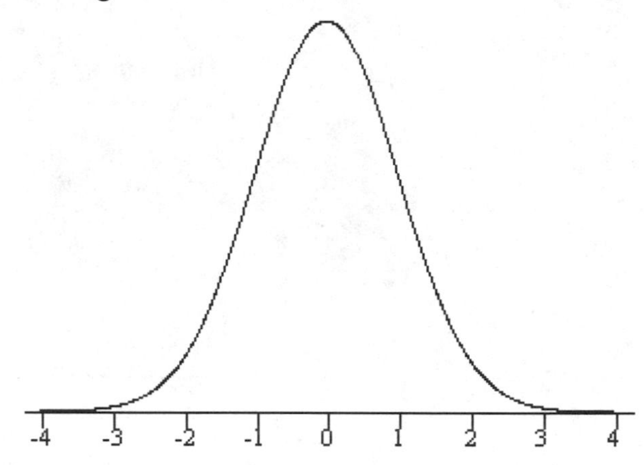

Remembering that the area under the whole curve is 1, how would you modify your answer from part (a)?

area =

(c) Try solving this part on your own. To begin, draw a picture of a normal curve and shade the region.

Now use the same line of reasoning as in part (b) to determine the area of your shaded region. Remember, you want to try to visualize your shaded region as a combination of areas of the form given in Table A.

(d) To test yourself, try this part on your own. It is a bit more complicated than the previous parts, but the same approach will work. Draw a picture and then try and express the desired area as the difference of two regions for which the areas can be found directly in Table A.

Exercise 1.123

KEY CONCEPTS: Finding the value z (the quantile) corresponding to a given area under a standard normal curve

The strategy used to solve this type of problem is the "reverse" of that used to solve Exercise 1.93. We again begin by drawing a picture of what we know; we know the area, but not z. For areas corresponding to those given in Table A we have a situation like the following.

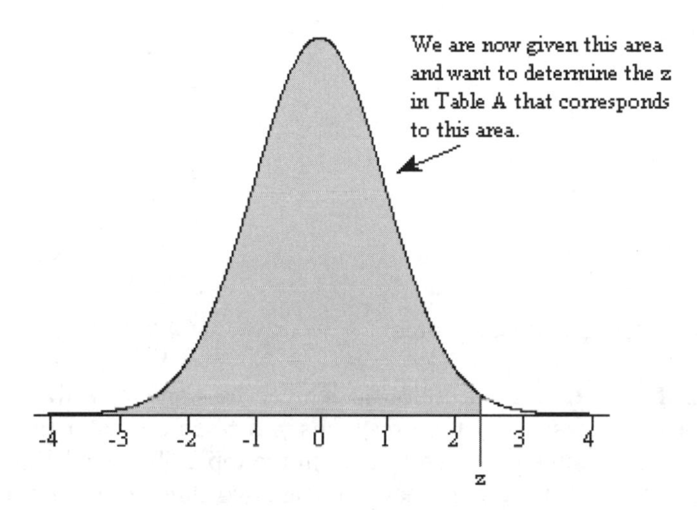
We are now given this area
and want to determine the z
in Table A that corresponds
to this area.
-4
-3
-2
-1
0
1
2
3
4
z

To determine z, we find the given value of the area in the body of Table A (or the entry in Table A closest to the given value of the area. We now look in the left margin of the table and across the top of the table to determine the value of z that corresponds to this area.

If we are given a more complicated area, we draw a picture and then determine from properties of the normal curve the area to the left of z. We then determine z as described above. The approach is illustrated in the solutions below.

(a) A picture of what we know is given below. Note that since the area given is larger than 0.5, we know z must be to the right of 0 (recall that the area to the left of 0 under a standard normal curve is 0.5).

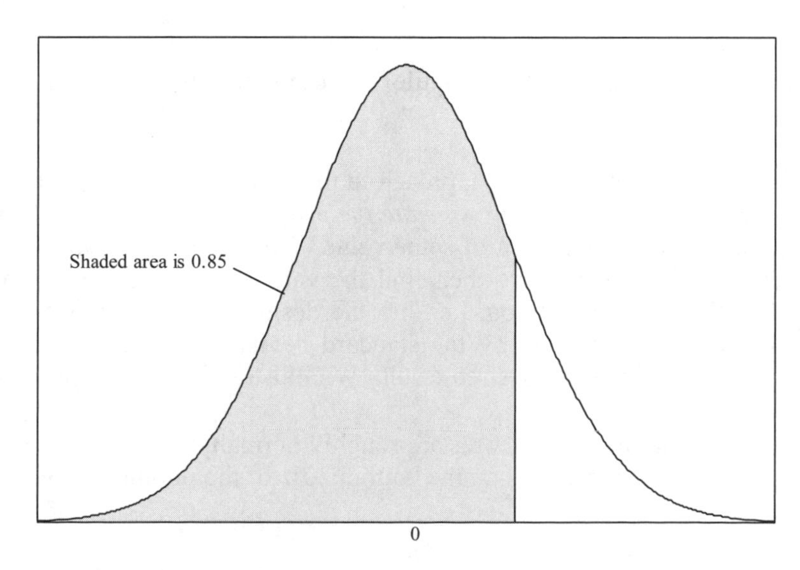
Shaded area is 0.85
0

We now turn to Table A and find the entry closest to 0.85. This entry is 0.8508. Locating the z-values in the left margin and top column corresponding to this entry, we see that the z that would give this area is 1.04.

(b) Try this part on your own. Begin by sketching a normal curve and the area you are given on the curve. On which side of 0 should z be located? Thinking about which side of zero a point lies on is a good way to make sure your answer makes sense.

Exercise 1.127

KEY CONCEPTS: Standardized scores, z-scores

To compare scores from two normal distributions, each can be standardized or converted into a z-score. For example, an ACT or an SAT score that corresponds to a z-score greater than 2 places either an ACT score in the top 2.5% of ACT scores or an SAT score in the top 2.5% of SAT scores. This is because in either case the z-score corresponds to a score that is at least two standard deviations above the mean of its distribution. Using the mean and standard deviation from each distribution, convert an ACT score of 16 and an ACT score of 670 to a z-score for each distribution.

ACT z-score = ________

SAT z-score = ________

Who has the higher score, Jacob or Emily? The z-score helps answer this.

Exercise 1.133

KEY CONCEPTS: Finding the value x (the quantile) corresponding to a given area under an arbitrary normal curve

To solve this problem, we must use a reverse approach to that which will be used in Exercise 1.136 (you may want to first solve Exercise 1.136). First we *state the problem*. To make use of Table A, we need to state the problem in terms of areas to the left of some value. Next, we *use the table*. To do so, we think of having standardized the problem and we then find the value z in the table for the standard normal distribution that satisfies the stated condition, i.e., has the desired area to the left of it. We next must *unstandardize* this z-value by multiplying by the standard deviation and then adding the mean to the result. This unstandardized value x is the desired result. We illustrate this strategy in the solution below.

State the problem. We know that the SAT scores are roughly normally distributed with $\mu = 1026$ and $\sigma = 209$. We want to find the score x that makes up the bottom 20% of the population of scores. This means that 20% of the population scores are less than x. We will need to first find the corresponding value z for the standard normal distribution. This is illustrated in the figure below.

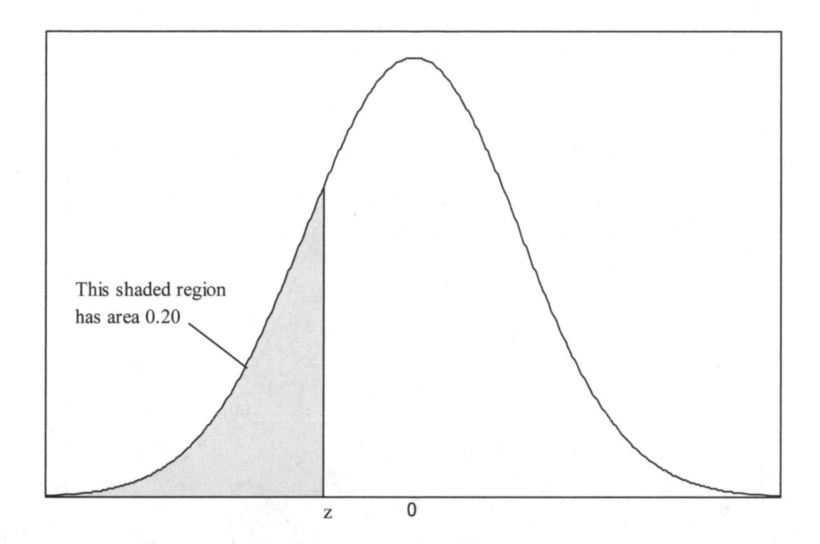

Use the table. The value z must have the property that the area to the left of it is 0.20. Areas to the left are the types of areas reported in Table A. Find the entry in the body of Table A that has a value closest to 0.20. Then find the value of z that yields this area.

Entry closest to 0.20 is _____________.

Value of z corresponding to the entry of 0.20 is _____________.

Unstandardize. We now must unstandardize z. Using the value of z corresponding to the entry of 0.20 in Table A, the unstandardized value is

$$x = (\text{standard deviation}) \times z + \text{mean} = 209 \times z + 1026 = \_\_\_\_\_\_\_\_$$

What SAT scores make up the bottom 20% of all scores?

Exercise 1.136

KEY CONCEPTS: Computing the area under an arbitrary normal curve

For these problems, we must first state the problem, then convert the question into one about a standard normal distribution. This involves standardizing the numerical conditions by subtracting the mean and dividing the result by the standard deviation. We then draw a picture of the desired area corresponding to these standardized conditions and compute the area as we did for the standard normal distribution, using Table A. This approach is illustrated in the solutions below.

(a) *State the problem.* Call the cholesterol level of a randomly chosen young woman X. The variable X has the $N(185, 39)$ distribution. We want the percent of young women with $X > 240$.

Standardize. We need to standardize the condition $X > 240$. We replace X by Z (we use Z to represent the standardized version of X) and standardize 240. Since we are told that the mean and standard deviation of cholesterol levels are 185 and 39, respectively, the standardized value (z-score) of 240 is (rounded to two decimal places)

$$z\text{-score of } 240 = (240 - 185)/39 = 1.41$$

In terms of a standard normal Z, the condition is $Z > 1.41$.

A picture of the desired area is

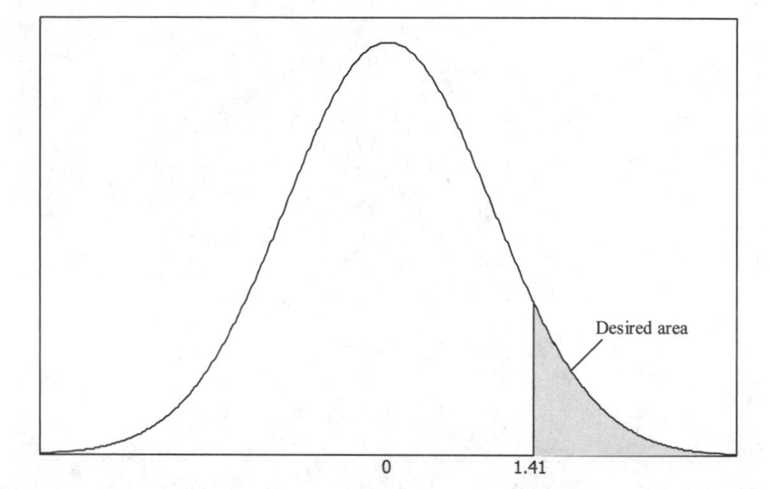

Use the table. The desired area is not of the form given in Table A. However, we note that the unshaded area to the left of 1.41 is of the form given in Table A and this area is 0.9207. Hence,

shaded area = total area under normal curve – unshaded area
= 1 – 0.9207 = 0.0793

Thus, the percent of young women whose cholesterol level X satisfies $X > 240$ is $0.0793 \times 100\% = 7.93\%$, or about 8%.

(b) Try this part on your own. First *state the problem.*

Next *standardize.* To do so, compute z-scores to convert the problem to a statement involving standardized values. In terms of z-scores, the condition of interest is

Standardized condition:

Sketch the standard normal curve below and shade the desired region on your curve.

Now *use the table.* Use Table A to compute the desired area. This will be the answer to the question.

Complete Solutions

Exercise 1.109

(a) The area of the density curve in this case is

area = length of base × height = ___4___ × ___h___

In order for this area to be 1, h must be 0.25.

Here is a graph of the density curve:

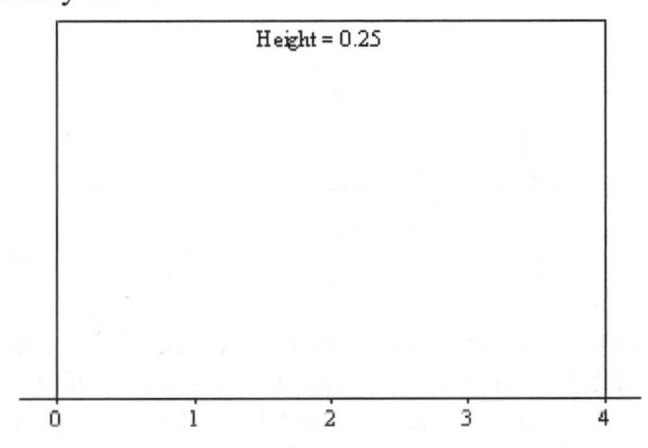

(b) The area of interest is the shaded region indicated below.

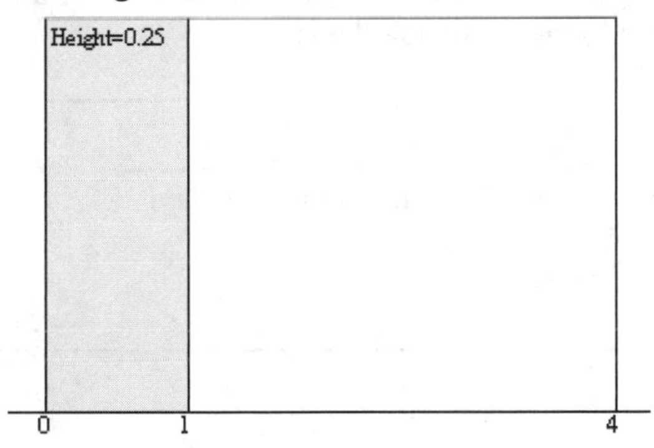

This shaded rectangular region has area = length of base × height = 1.0 × 0.25 = 0.25.

(c) The area of interest is the shaded region indicated below.

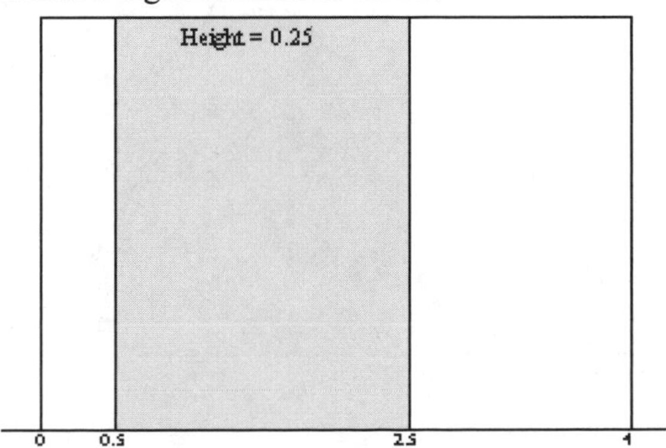

This rectangular region has area = length of base × height = 2.0 × 0.25 = 0.5.

Exercise 1.115

(a) The shaded region lies between 327 and 345 days.

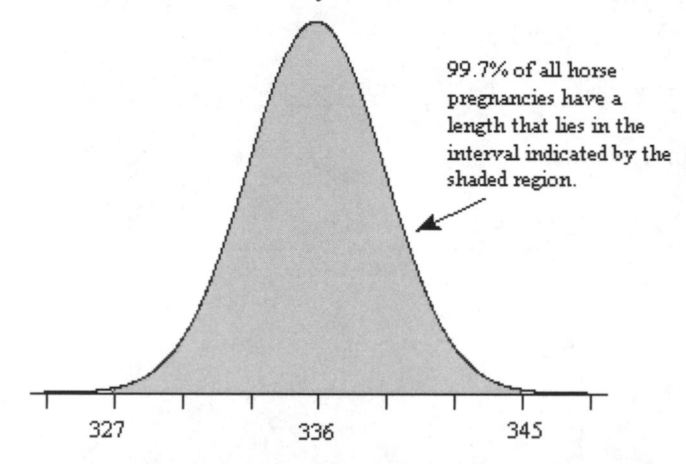
99.7% of all horse pregnancies have a length that lies in the interval indicated by the shaded region.
327
336
345

(b) Refer to the figure in part (b) of this exercise that is provided in the Guided Solutions. If the shaded region gives the middle 68% of the area, then the two unshaded regions must account for the remaining 32%. Since the normal curve is symmetric, each of the two unshaded regions must have the same area and each must account for half of the 32%. Hence, each of the unshaded regions accounts for 16% of the area. The rightmost of these regions accounts for the longest 16% of all pregnancies. We conclude that 16% of all horse pregnancies are longer than 339 days.

Exercise 1.121

(a) A complete solution was provided in the Guided Solutions.

(b) The desired area is indicated below.

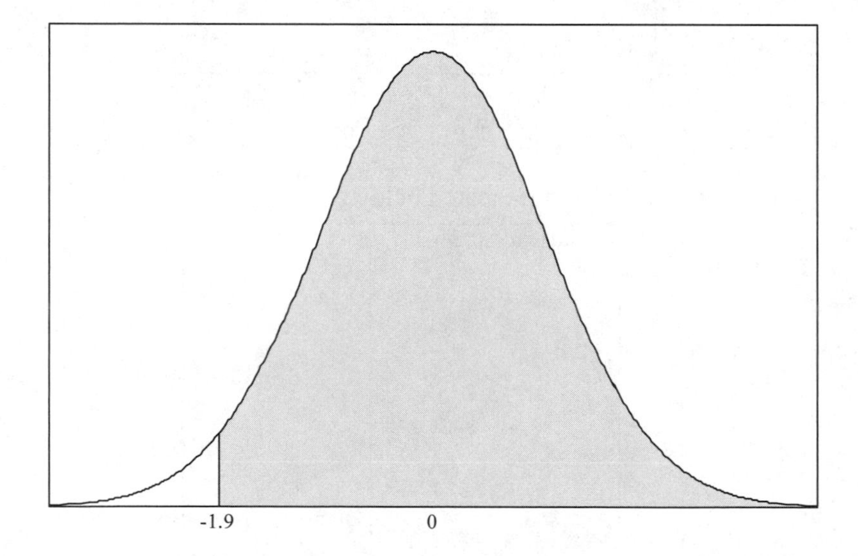
-1.9
0

This is not of the form for which Table A can be used directly. However, the unshaded area to the left of –1.90 is of the form needed for Table A. In fact, we found the area of the unshaded portion in part (a).

We notice that the shaded area can be visualized as what is left after deleting the unshaded area from the total area under the normal curve.

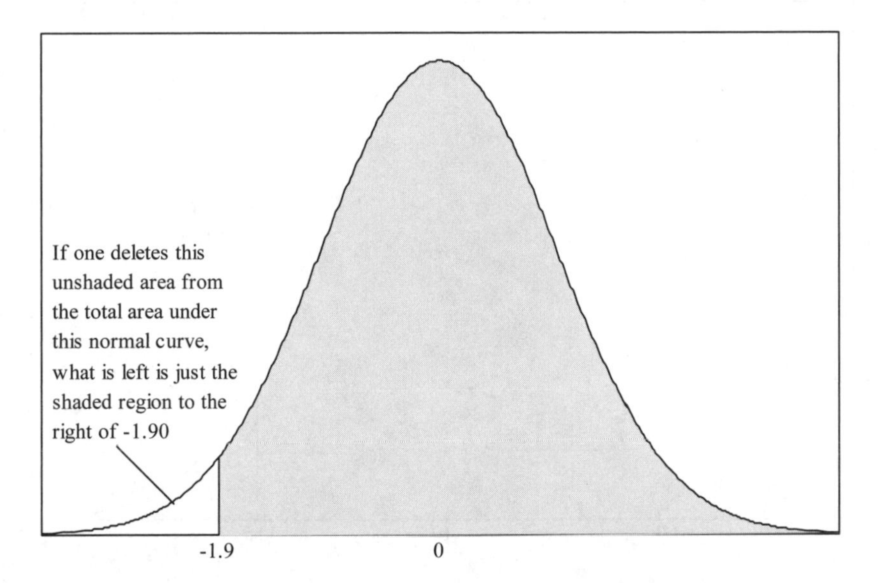

Since the total area under a normal curve is 1, we have

shaded area = total area under normal curve – area of unshaded portion
= 1 – 0.0287 = 0.9713

Thus, the desired proportion is 0.9713.

(c) The desired area is indicated in the figure below.

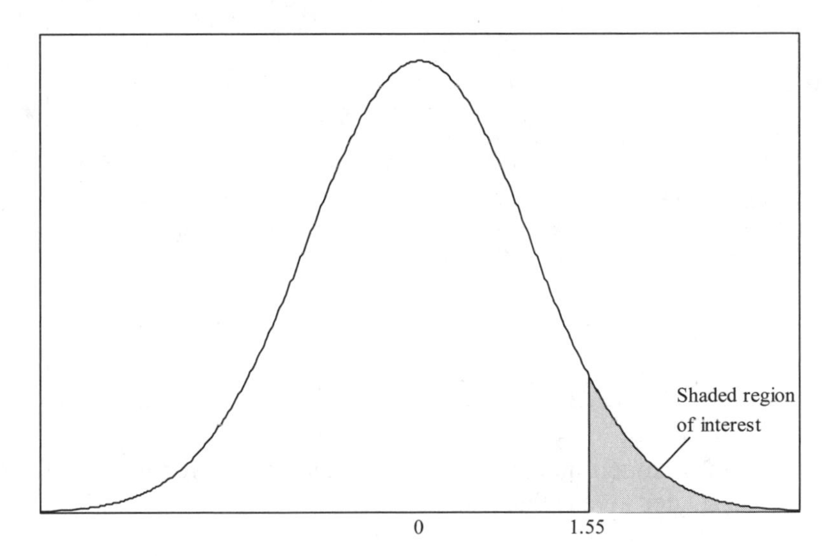

This is just like part (b). The unshaded area to the left of 1.55 can be found in Table A and is 0.9394. Thus,

shaded area = total area under normal curve – area of unshaded portion
= 1 – 0.9394 = 0.0606

This is the desired proportion.

(d) We begin with a picture of the desired area.

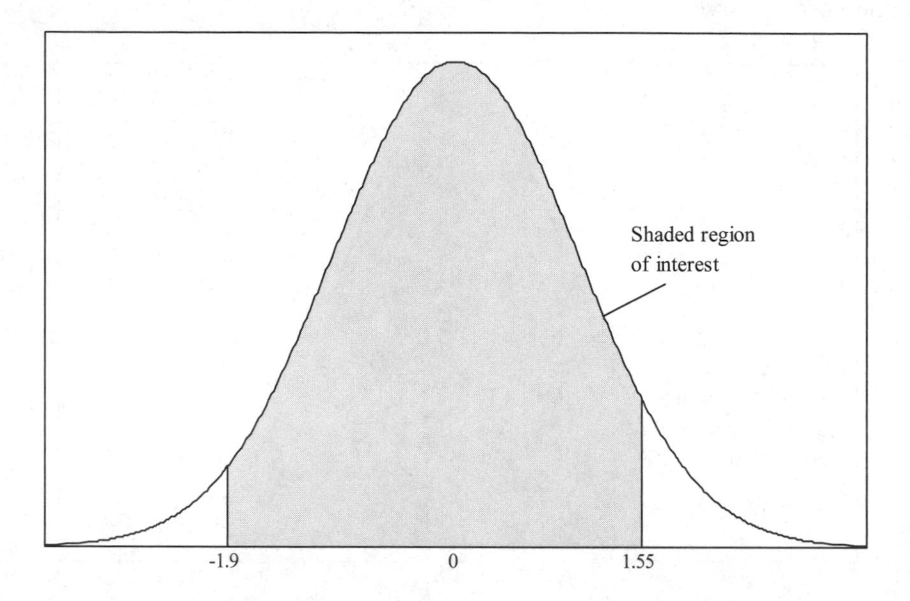

The shaded region is a bit more complicated than in the previous parts: however, the same strategy still works. We note that the shaded region is obtained by removing the area to the left of –1.90 from all the area to the left of 1.55.

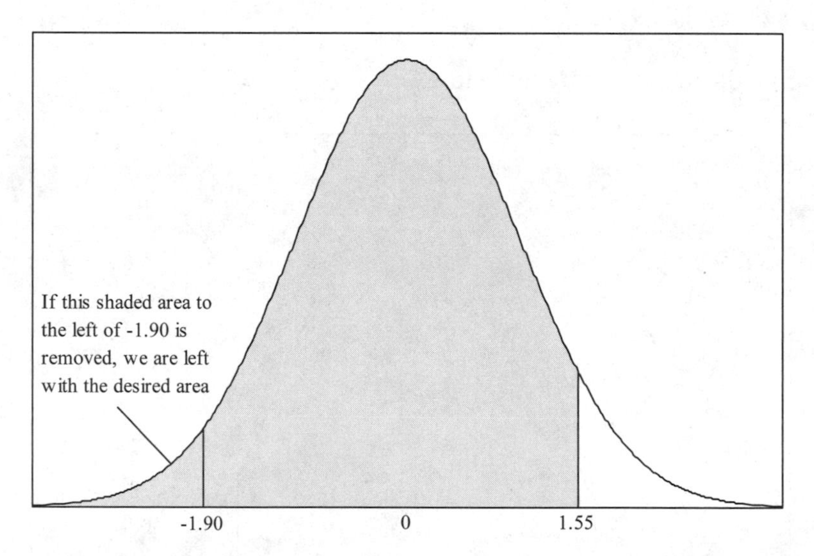

The area to the left of –1.90 is found in Table A to be 0.0287. The area to the left of 1.55 is found in Table A to be 0.9394. The shaded area is thus

$$\begin{aligned}\text{shaded area} &= \text{area to left of } 1.55 - \text{area to left of } -1.90 \\ &= 0.9394 - 0.0287 \\ &= 0.9107\end{aligned}$$

This is the desired proportion.

Exercise 1.123

(a) A complete solution was given in the Guided Solutions.

(b) A picture of what we know is given below. Note that since the area to the right of 0 under a standard normal curve is 0.5, we know that z must be located to the right of 0.

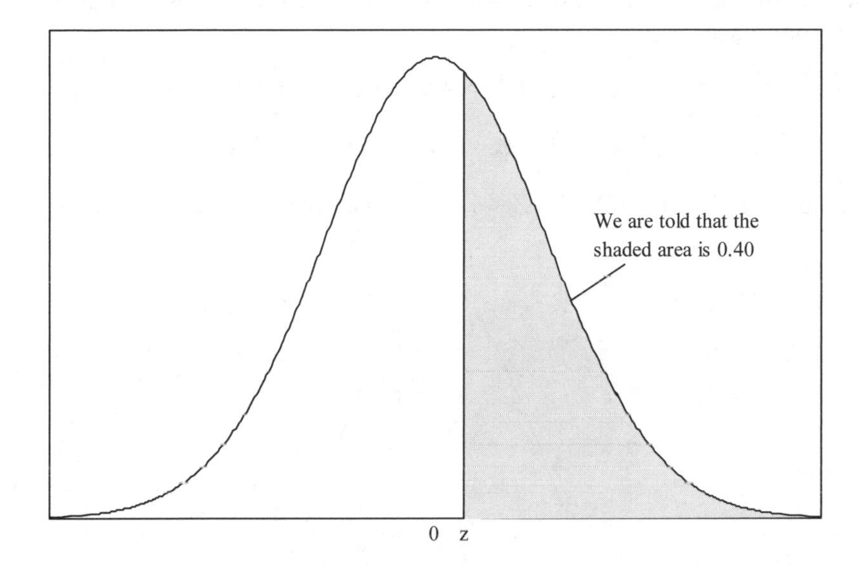
We are told that the
shaded area is 0.40
0
z

The shaded area is not of the form used in Table A. However, we note that the unshaded area to the left of z is of the correct form. Since the total area under a normal curve is 1, this unshaded area must be $1 - 0.40 = 0.60$. Hence, z has the property that the area to the left of z must be 0.60. We locate the entry in Table A closest to 0.60. This entry is 0.5987. The z corresponding to this entry is 0.25.

Exercise 1.127

ACT $\quad z\text{-score} = \dfrac{17 - 20.8}{4.8} = -0.79$

SAT $\quad z\text{-score} = \dfrac{680 - 1026}{209} = -1.66$

Both Jacob and Emily have scored below the mean, putting them in the lower half of the scores. But Emily's SAT score of 680 is 1.66 standard deviations below the mean SAT score, while Jacob's ACT score is 0.79 standard deviations below the mean ACT score, so Jacob has the higher score.

Exercise 1.133

We find the entry in the body of Table A that has a value closest to 0.20. This entry is 0.2005. The value of z that yields this area is seen, from Table A, to be –0.84.

Unstandardize. We now must unstandardize z. The unstandardized value is

$$x = (\text{standard deviation}) \times z + \text{mean} = 209 \times z + 1026 = (209 \times (-0.84)) + 1026 = 850.44$$

SAT scores below 850 make up the bottom 20% of all scores.

Exercise 1.136

(a) A complete solution was given in the Guided Solutions.

(b) *State the problem.* The problem is to find the percent of young women with $200 < X < 240$. *Standardize.* We need to first standardize the condition $200 < X < 240$. We replace X by Z (we use Z to represent the standardized version of X) and standardize 200 and 240. Since we are told that the mean and standard deviation are 185 and 39, respectively, the standardized values (z-scores) of 200 and 240 are (rounded to two decimal places)

$$z\text{-score of } 200 = (200 - 185)/39 = 0.38$$

$$z\text{-score of } 240 = (240 - 185)/39 = 1.41$$

Our condition in standardized form is $0.38 < Z < 1.41$. A picture of the desired area is

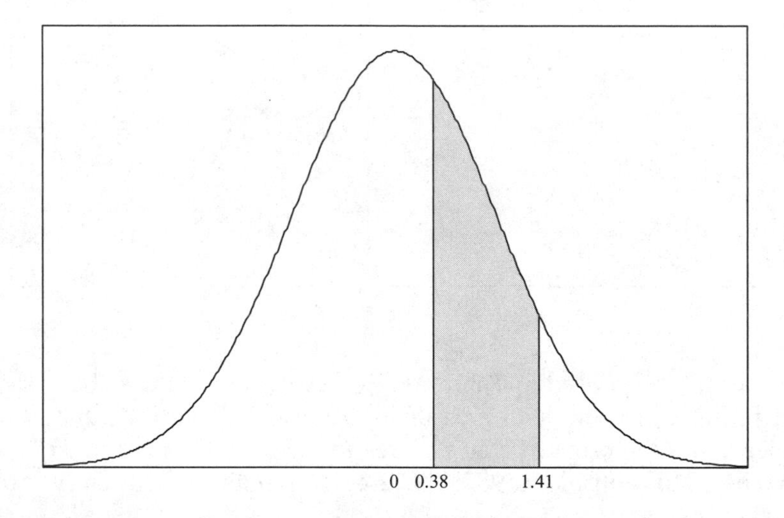

Use the table. This area can be found by determining the area to the left of 1.41 and the area to the left of 0.38, and then subtracting the area to the left of 0.38 from the area to the left of 1.41. From Table A

$$\text{area to the left of } 1.41 = 0.9207$$

$$\text{area to the left of } 0.38 = 0.6480$$

and the desired difference is $0.9207 - 0.6480 = 0.2727$. Thus, the percent of young women whose cholesterol level X satisfies $200 < X < 240$ is $0.2727 \times 100\% = 27.27\%$.

CHAPTER 2

LOOKING AT DATA— RELATIONSHIPS

SECTION 2.1

OVERVIEW

The first chapter provides the tools to explore several types of variables one by one, but in most instances the data of interest are a collection of variables that may exhibit relationships among themselves. Typically, these relationships are more interesting than the behavior of the variables individually. If we think that one of the variables, x, may explain or even cause changes in another variable, y, we call x an **explanatory variable** and y a **response variable.**

The first tool we consider for examining the relationship between variables is the **scatterplot.** Scatterplots show us two quantitative variables at a time, such as the weight of a car and its MPG (miles per gallon). Using colors or different symbols, we can add information to the plot about a third variable that is categorical in nature. For example, if in our plot we wanted to distinguish between cars with manual or automatic transmissions, we might use a circle to plot the cars with manual transmissions and a cross to plot the cars with automatic transmissions.

When drawing a scatterplot, we need to pick one variable to be on the horizontal axis and the other to be on the vertical axis. When there is a response variable and an explanatory variable, the explanatory variable is always placed on the horizontal axis. In cases where there is no explanatory-response variable distinction, either variable can go on the horizontal axis. After drawing the scatterplot by hand or using a computer, the scatterplot should be examined for an **overall pattern** that may tell us about any relationship between the variables, as well as for **deviations** from this overall pattern. You should be looking for the **direction, form,** and **strength** of the overall pattern. In terms of direction, **positive association** occurs when the variables both take on high values together, while **negative association** occurs if one variable takes on high values when the other takes on low values. In many cases, when an association is present, the variables appear to have a **linear relationship.** The plotted values seem to cluster around a line. If the line slopes up to the right, the association is positive; and if the line slopes down to the right, the association is negative. As always, look for **outliers.** The outlier may be far away in terms of the horizontal variable or the vertical variable or far away from the overall pattern of the relationship.

GUIDED SOLUTIONS

Exercise 2.9

KEY CONCEPTS: Explanatory and response variables

(a) When examining the relationship between two variables, if you hope to show that one variable can be used to explain variation in the other, remember that the response variable measures the outcome of the study, while the explanatory variable explains changes in the response variable. When you just want to explore the relationship between two variables such as scores on the math and verbal SAT, then the explanatory-response variable distinction is not important.

In this case, it seems reasonable to view the age of a child as explaining their weight from birth to ten years. Thus, the weight is the response and the age is the explanatory variable. Now try the other parts on your own.

(b)

(c)

(d)

(e)

Exercise 2.15

KEY CONCEPTS: Drawing and interpreting a scatterplot

(a) Suppose the percent of returning males had decreased slightly, but we had made the response variable the number of returning males. What would the relationship between pairs and number be?

(b) When drawing a scatterplot, we first need to pick one variable (the explanatory variable) to be on the horizontal axis and the other (the response) to be on the vertical axis. In this data set we are interested in the "effect" of the number of pairs on the percent of males who return. So the number of pairs is the explanatory variable and the percent of males who return is the response. In the figure on the left at the top of the next page, we have plotted the first observation corresponding to 28 pairs and a percent of 82. Complete this scatterplot. Although you will generally draw scatterplots on the computer, drawing a small one like this by hand ensures that you understand what the points represent. In the figure on the right, plot the mean responses for each of six values of the explanatory variable. That is, for 29 pairs just plot the average percent for these three observations, rather than the three distinct responses. For 38 pairs, plot the average of the two responses. Plotting the average will reduce some of the variability and will emphasize the pattern.

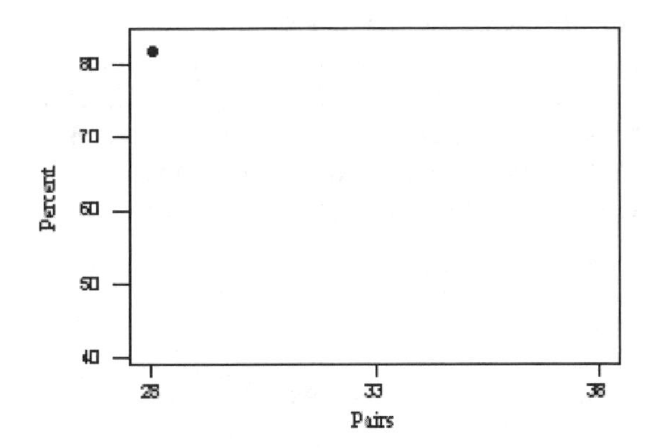

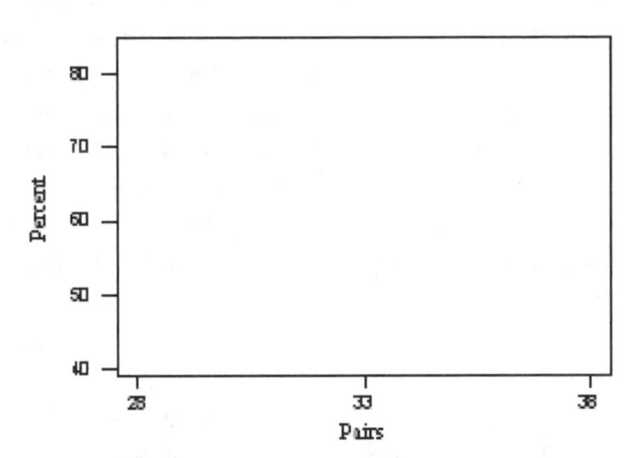

(c) Describe the pattern. Is the association positive or negative? Do years with higher numbers of pairs tend to have higher or lower percents of males returning? Do the data support the theory that a smaller percent of birds survive following a successful breeding season?

Exercise 2.21

KEY CONCEPTS: Drawing and interpreting a scatterplot, adding a categorical variable to a scatterplot

(a) When drawing a scatterplot, we first need to pick one variable (the explanatory variable) to be on the horizontal axis and the other (the response) to be on the vertical axis. In this data set we are interested in the "effect" of lean body mass on metabolic rate. So lean body mass is the explanatory variable and metabolic rate is the response variable in the following plot. Although you will generally draw scatterplots on the computer, drawing a small one like this by hand ensures that you understand what the points represent.

The scatterplot with the points for the females is given below. Add the data for the males to this plot using a different color or plotting symbol.

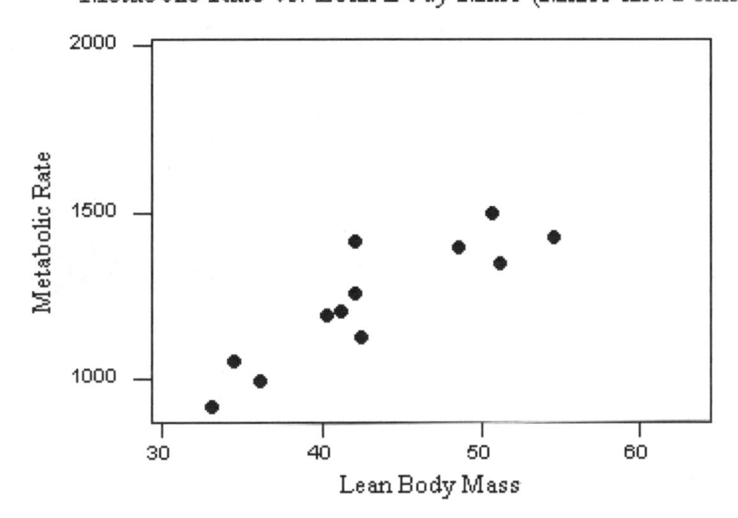

(b) Here are some guidelines for examining scatterplots: Do the data show any association? **Positive association** is when the variables both take on high values together. **Negative association** is when one variable takes on high values and the other takes on low values. If the plotted values seem to form a line, the variables may have a **linear relationship**. If the line slopes up to the right, the association is positive. If the line slopes down to the right, the association is negative. Are there any **clusters** of data? Clusters are distinct groups of observations. As always, look for outliers. An **outlier** may be far away in terms of the horizontal variable or the vertical variable or far away from the overall pattern of the relationship.

For our plot, is the overall association positive or negative? What is the overall form of the relationship? How strong is the overall relationship?

Is the pattern of the relationship for the men similar to that for the female subjects? If not, how do the male subjects as a group differ from the female subjects as a group?

Exercise 2.26

KEY CONCEPTS: Categorical variables in scatterplots

(a) In the scatterplot, there should be four observations above each of the levels of nematode count. After adding these points to the graph below, compute the mean of the four observations at each level of nematode count, and put each mean on the graph. Then connect the four means.

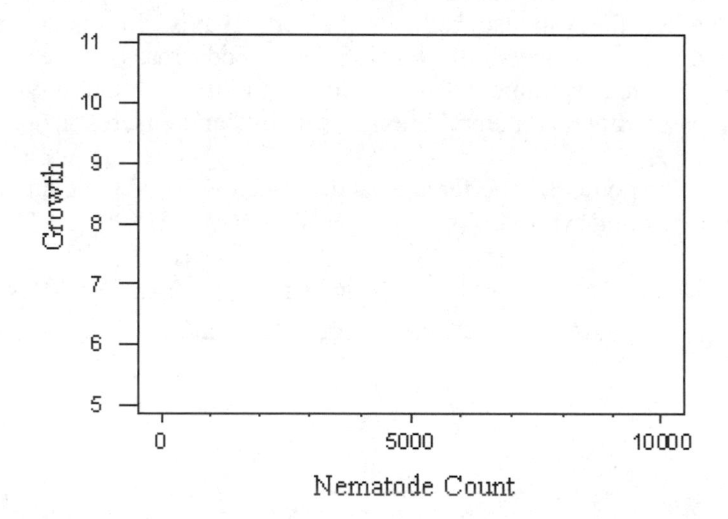

(b) What sorts of changes do you see in the means as the nematode count increases?

COMPLETE SOLUTIONS

Exercise 2.9

(a) A complete solution was provided in the Guided Solutions.

(b) We would probably simply want to explore the relationship between high school English grades and high school math grades.

(c) We would probably view the number of bedrooms as explaining the apartment rental price. Thus, the response is the apartment rental price, and the explanatory variable is the number of bedrooms.

(d) We would probably view the amount of sugar added to the cup of coffee as explaining how sweet the coffee tastes. Thus, the response is how sweet the coffee tastes, and the explanatory variable is the amount of sugar added to the cup of coffee.

(e) We would probably simply want to explore the relationship between the student evaluation scores for an instructor and the student evaluation scores for the course.

Exercise 2.15

(a) If there were more pairs, you would expect the number of males returning to increase. If a smaller *percent* returned, the *number* returning could still be higher than when there were fewer pairs even though the percent returning decreased. The hypothesis is that the *percent* that survive to breed again is lower following a successful breeding season, and this should be the response.

(b) The graph on the right, which plots the mean response, emphasizes the pattern. For 29 pairs we have plotted the average response of the three percents, $(83 + 70 + 61)/3 = 71.33$ and similarly for 38 pairs.

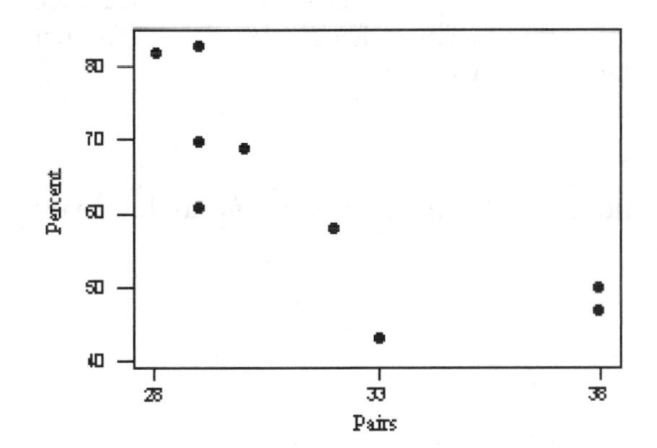

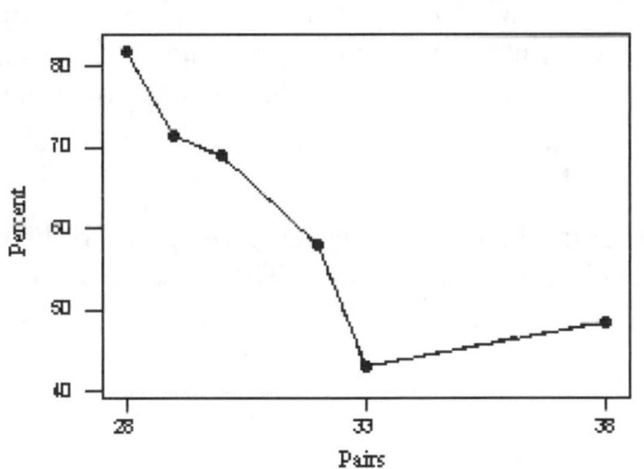

(c) The pattern suggests that the percent returning tends to decrease as the number of pairs for the preceding season increases. The decrease seems fairly pronounced between 28 and 33 pairs and then seems to level off. Don't over-interpret the plot by concluding that after 33 pairs the percent increases again. The slight increase could be due to a single low observation at 33, not a real pattern.

Exercise 2.21

(a) We add the men to the plot. Men are indicated by the +'s in the plot.

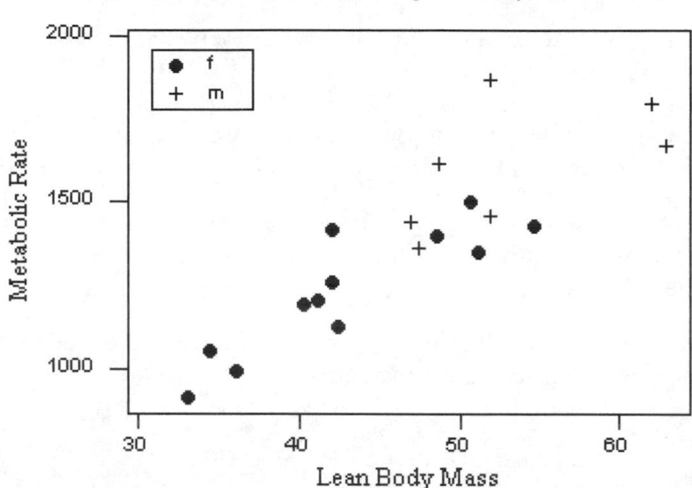

(b) As lean body mass increases, or as you move from left to right across the horizontal axis in the scatterplot, the points in the plot tend to rise. This indicates that the association between the variables is positive. The form of the relationship appears to be linear since a straight line seems to be a reasonable approximation to the overall trend in the plot. The relationship is not perfect, but it appears to be moderately strong.

The pattern of the relationship is roughly the same for men and women. The strength of the relationship for females appears to be slightly stronger than for males. The most striking difference between the points corresponding to male and female subjects is that the men are clustered in the upper right of the plot. This is not surprising, since men tend to be larger than women.

Exercise 2.26

(a) In the graph below, the circles correspond to the observations and the pluses correspond to the means at each level of nematode count. The pluses are connected.

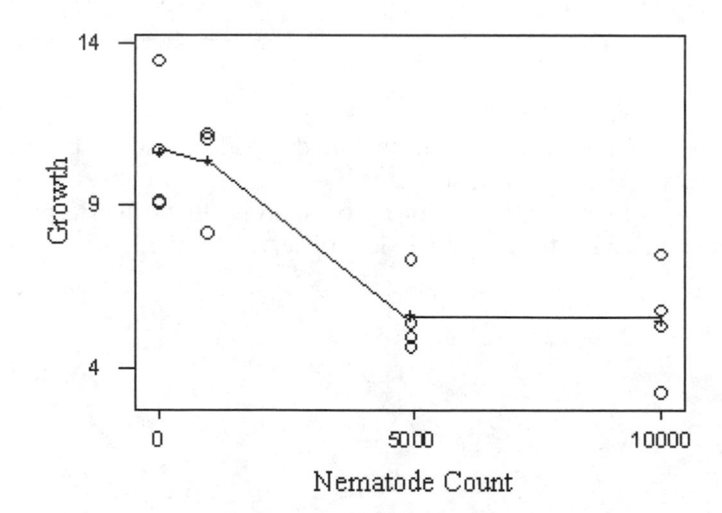

(b) The level of growth may decrease slightly when going from 0 to a count of 1000, but then it drops off considerably by 5000 and seems to stay at that level until 10,000. In making these statements we are making some assumptions about the average growth between the levels of nematode counts in our experiments (interpolating). We claimed that the growth level did not change between 5000 and 10,000, but we have no data between those values. At what value of count between 1000 and 5000 does the drop in level of growth occur, or is it dropping steadily over this range of counts? It would be helpful to have had an observation at 3000, as the data suggests there was a fairly sharp decrease in growth at some nematode count between 1000 and 5000, but we can't say more.

SECTION 2.2

OVERVIEW

Scatterplots provide a visual tool for looking at the relationship between two variables. Unfortunately, our eyes are not good tools for judging the strength of the relationship. Changes in the scale or the amount of white space in the graph can easily affect our judgment as to the strength of the relationship. **Correlation** is a numerical measure we will use to show the strength of **linear association**.

The correlation can be calculated using the formula

$$r = \frac{1}{n-1}\sum\left(\frac{x_i - \bar{x}}{s_x}\right)\left(\frac{y_i - \bar{y}}{s_y}\right)$$

where $\bar{x}$ and $\bar{y}$ are the respective means for the two variables X and Y, and s_x and s_y are their respective standard deviations. In practice, you will probably compute the value of r using computer software or a calculator that finds r from entering the values of the x's and y's. When computing a correlation coefficient there is no need to distinguish between the explanatory and response variables, even in cases where this distinction exists. The value of r will not change if we switch x and y.

When r is positive it means that there is a positive linear association between the variables, and when it is negative there is a negative linear association. The value of r is always between 1 and –1. Values close to 1 or –1 show a strong association while values near 0 show a weak association. As with means and standard deviations, the value of r is strongly affected by outliers. Their presence can make the correlation much different than what it might be with the outlier removed. Finally, remember that the correlation is a measure of straight line association. There are many other types of association between two variables, but these patterns will not be captured by the correlation coefficient.

GUIDED SOLUTIONS

Exercise 2.29

KEY CONCEPTS: Interpreting and computing the correlation coefficient

(a) In this data set we are interested in the "effect" of the price of coffee on the rate of deforestation. So the price of coffee is the explanatory variable. Make a scatterplot of the data on the plot on the next page. What kind of pattern is shown in your plot?

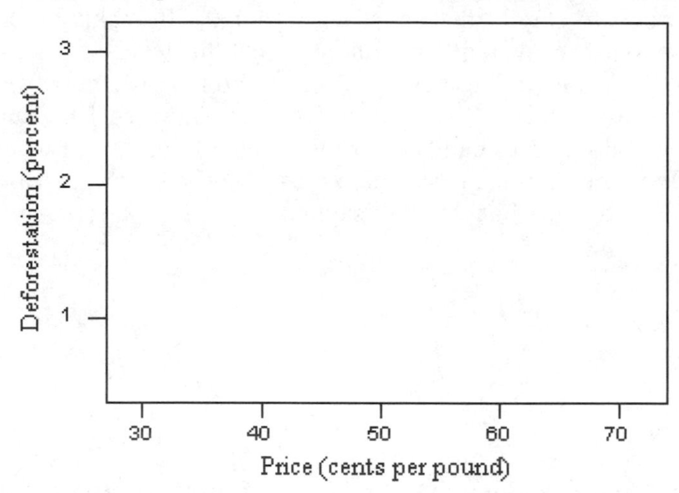

(b) Find the correlation r step-by-step. The means and standard deviations of the variables are given below.

$\bar{x} = 50.00$ $s_x = 16.32$

$\bar{y} = 1.738$ $s_y = 0.928$

The step-by-step calculations for the correlation r can be organized in the following table. Use the means and standard deviations given above to complete the table.

x	$\frac{x_i - \bar{x}}{s_x}$	y	$\frac{y_i - \bar{y}}{s_y}$	$\left(\frac{x_i - \bar{x}}{s_x}\right)\left(\frac{y_i - \bar{y}}{s_y}\right)$
29		0.49		
40		1.59		
54		1.69		
55		1.82		
72		3.10		

What is the sum of the values in the last column above? The correlation is this sum divided by $n - 1$.

Sum of the values in the last column = __________ r = ____________

(c) Use a calculator or software to determine r. Compare with your answer in (b).

Exercise 2.37

KEY CONCEPTS: Linear transformations, effect on correlation

Is the transformation from dollars to euros a linear transformation? What is the effect of a linear transformation on the correlation coefficient?

Exercise 2.45

KEY CONCEPTS: Computing the correlation coefficient, the effect of outliers

(a) Make your scatterplot on the axes provided.

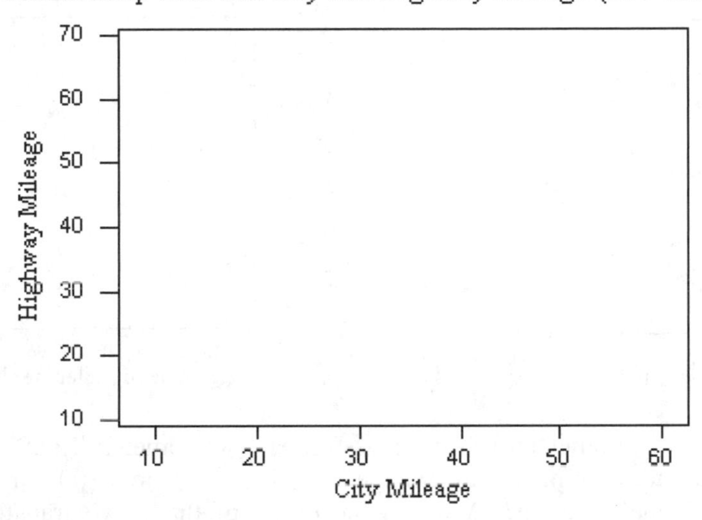

Does the Insight extend the linear pattern of the other cars, or is it far from the line they form?

(b) Compute the correlations and enter the results in the space provided. Use statistical software if available. If you are computing the correlations by hand, you may find it useful to organize your calculations as we did in Exercise 2.29.

Correlation with all observations = ___________________

Correlation without the Insight = ___________________

Explain the difference in the two values based on your answer to (a).

Exercise 2.49

KEY CONCEPTS: Interpreting the correlation coefficient

The problem is that a correlation close to zero and the quote "good researchers tend to be poor teachers, and vice versa" are not the same. What does a correlation close to zero mean? What would be true about the correlation if "good researchers tend to be poor teachers, and vice versa"?

Exercise 2.52

KEY CONCEPTS: Effect of transformations on the correlation coefficient

(a) The first plot is speed (km/h) vs. fuel (liters/100 km) and the second plot is also speed (miles/hour) vs. fuel (gallons/mile). They are virtually identical except for the labeling of the axes, which corresponds to the units in which the variables are measured.

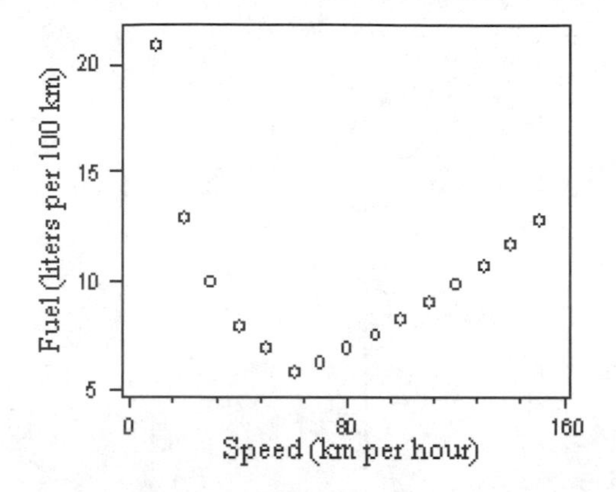

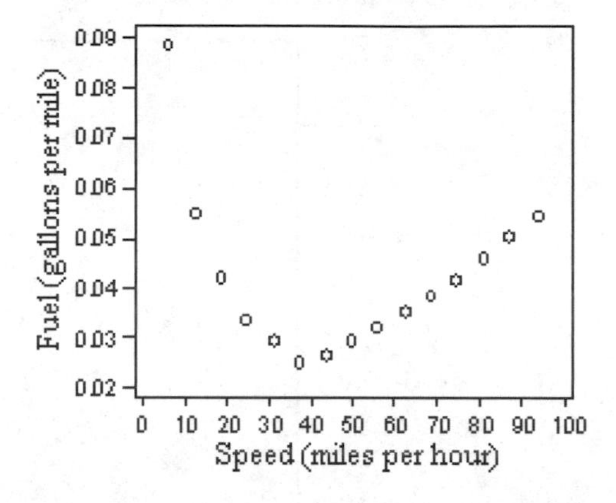

The transformation to change kilometers to miles is miles = kilometers/1.609 and is a linear transformation. What is the transformation to change fuel used in liters/100 km to fuel used in gallons/mile? Is it a linear transformation? What is the effect of these two transformations on the numerical value of the correlation coefficient? (**Note:** Since the relationship between speed and fuel consumption is not linear, do you think the correlation coefficient is a good summary of the "strength" of the relationship?)

(b) The plot below is of fuel consumption in miles per gallon rather than in gallons per mile, and speed is in miles per hour.

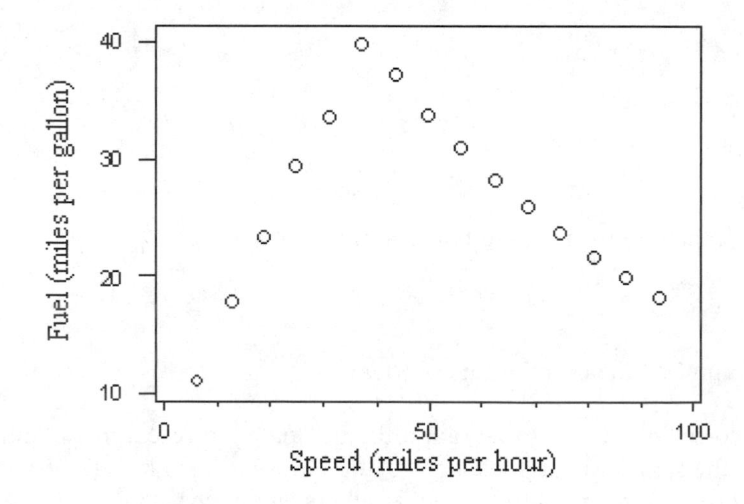

What is the transformation to change fuel used in gallons per mile to miles per gallon? What is the transformation to change liters per 100 km to miles per gallon? Is this transformation linear? What is the effect on the correlation coefficient? **Note:** The relationship is still not linear, so the correlation coefficient is again not a very good summary measure.

COMPLETE SOLUTIONS

Exercise 2.29

(a) There is an increasing trend. The strong linear trend is due to the data points at the very low price of coffee (29 cents) and the very high price of coffee (72 cents). Without these points, the correlation would not be nearly as high.

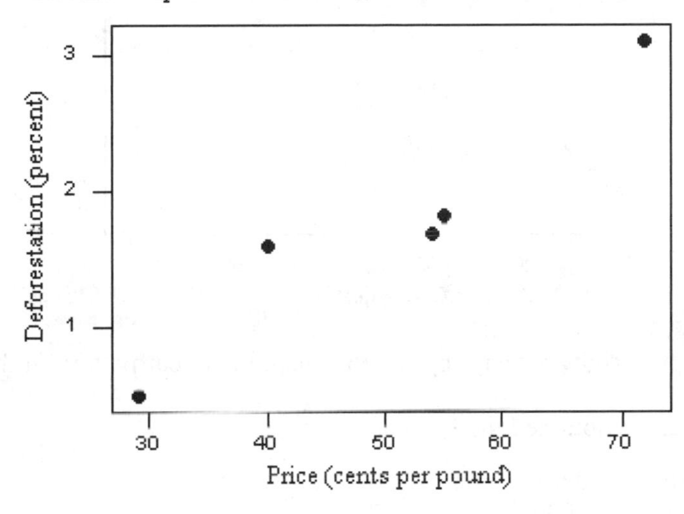

(b) The table is completed below.

x	$\frac{x_i - \bar{x}}{s_x}$	y	$\frac{y_i - \bar{y}}{s_y}$	$\left(\frac{x_i - \bar{x}}{s_x}\right)\left(\frac{y_i - \bar{y}}{s_y}\right)$
29	-1.28676	0.49	-1.34483	1.73048
40	-0.61275	1.59	-0.15948	0.09772
54	0.24510	1.69	-0.05172	-0.01268
55	0.30637	1.82	0.08836	0.02707
72	1.34804	3.10	1.46767	1.97848

The sum of the values in the last column is 3.82107. The correlation is this sum divided by $n - 1 = 4$.

$$r = 3.82107/4 = 0.955.$$

(c) Software gives $r = 0.955$, which agrees with (b). When doing the calculations by hand, be careful about rounding off numbers in the intermediate stages of the calculations.

Exercise 2.37

At the time of this writing a euro was worth \$1.47, or euro = 1.47 × dollars, which is a linear transformation with $a = 0$ and $b = 1.47$. The value of the correlation is unchanged under linear transformations of the variables.

Exercise 2.45

(a)

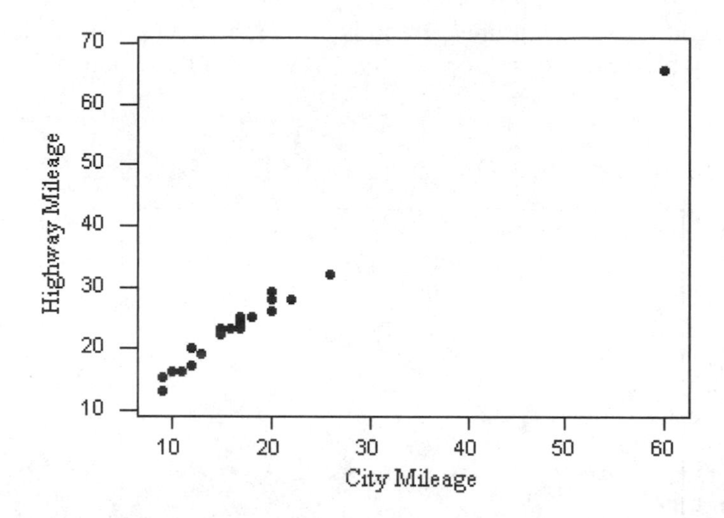

The Insight (outlier in the upper-right corner) appears to extend the linear pattern of the other cars.

(b) Using statistical software, we obtained the following.

Correlation with all observations = 0.993
Correlation without the Insight = 0.976

The correlation is closer to 1 when we include the Insight. As we noted in (a), the Insight appears to extend the linear pattern of the other cars. In so doing, it actually strengthens the visual impression of the linear pattern because it is far from the other points in the plot.

Exercise 2.49

If the correlation were close to zero, there would be no particular linear relationship. Good researchers would be just as likely as bad researchers to be good or bad teachers. The statement that "good researchers tend to be poor teachers, and vice versa" implies that the correlation is negative, not zero.

Exercise 2.52

(a) The transformation here is $\text{fuel}_{(\text{gallons per mile})} = (1.609/378.5) \times \text{fuel}_{(\text{liters per 100 km})}$ and is linear. Since the transformations of x and y are both linear, the correlation coefficient is the same for the original and transformed variables. The value of the correlation coefficient is –0.172. Since the relationship is clearly nonlinear, and the correlation only measures the strength of a linear relationship, it would be a mistake to interpret this as a negative relationship between fuel consumption and speed.

(b) To go from gallons per mile to miles per gallon we need to take the reciprocal miles per gallon = 1/gallons per mile, or

$$\text{MPG} = 378.5/(1.609 \times \text{fuel}_{(\text{liters per 100 km})})$$

This is not a linear transformation, so the numerical value of the correlation coefficient will change. The value of the correlation coefficient is now –0.043. A value near zero would suggest a weak relationship—but that refers to a weak **linear** relationship. There is clearly a very strong relationship between speed and miles per gallon, but it is nonlinear.

SECTION 2.3

OVERVIEW

If a scatterplot shows a linear relationship that is moderately strong as measured by the correlation, we can draw a line on the scatterplot to summarize the relationship. In the case where there is a response and an explanatory variable, the **least-squares regression line** often provides a good summary of this relationship. A straight line relating y to x has the form

$$y = a + bx$$

where b is the **slope** of the line and a is the **intercept.** The slope tells us the change in y corresponding to a one-unit increase in x. The intercept tells us the value of y when x is 0; this has no practical meaning unless 0 is a value that x takes in practice.

The least-squares regression line is the straight line $\hat{y} = a + bx$, which minimizes the sum of the squares of the vertical distances between the line and the observed values y. The formula for the slope of the least-squares line is

$$b = r\frac{s_y}{s_x}$$

and for the intercept it is $a = \bar{y} - b\bar{x}$, where $\bar{x}$ and $\bar{y}$ are the means of the x and y variables, s_x and s_y are their respective standard deviations, and r is the value of the correlation coefficient. Typically, the equation of the least-squares regression line is obtained by computer software or a calculator with a regression function.

Regression can be used to predict the value of y for any value of x. Just substitute the value of x into the equation of the least-squares regression line to get the predicted value for y. Predicting values of y for x-values in the range of those x's we observed is called interpolation and is fine to do. However, be careful about **extrapolation** (using the line for prediction beyond the range of x-values covered by the data). Extrapolation may lead to misleading results if the pattern found in the range of the data does not continue outside the range.

Correlation and regression are clearly related; as can be seen from the equation for the slope, b. However, the more important connection is how r^2, the square of the correlation coefficient, measures the strength of the regression. The square of r tells us the fraction of the variation in y that is explained by the regression of y on x. The closer r^2 is to 1, the better the regression describes the connection between x and y.

GUIDED SOLUTIONS

Exercise 2.63

KEY CONCEPTS: Least-squares regression, interpreting the slope, extrapolation

(a) Which quantity in the regression equation can be used to answer this question? Review the definitions of the slope and the intercept.

(b) To predict the value of y for any value of x, just substitute the value of x into the equation of the least-squares regression line to get the predicted value for y. In this case, $x = 1880$. What is the predicted value of y? Why is this extrapolation?

(c) Use Table 1.4 to find the observed value of y in 1990 and the regression equation to compute the predicted value. From these two values, you can compute

Error = observed y – predicted y = ____________________

(d) Would the great floods cause the observed values to lie above or below the regression line? Examination of the plot should reveal both of these years, even on the plot of annual water discharged.

Exercise 2.67

KEY CONCEPTS: Scatterplots, least-squares regression, prediction

(a) Make your scatterplot on the axes below. We have used software to obtain the equation of the least-squares regression line. The equation is

```
Activity = -0.126 + 0.0608 Distress
```

Once you have completed the scatterplot, add the regression line to your plot.

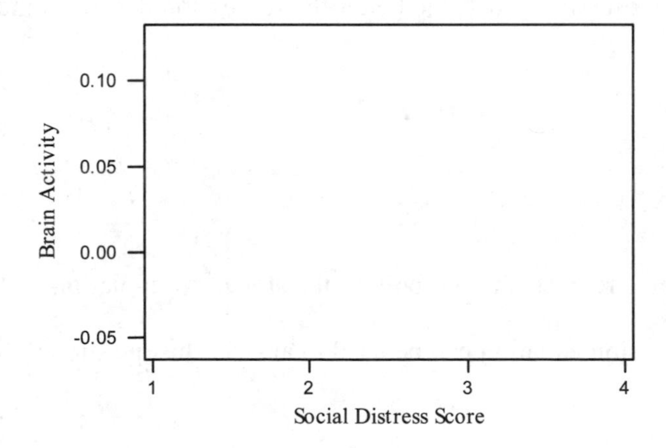

(b) On your graph in (a), fill in the "up and over" lines to predict brain activity for a social distress score of 2.0. Now use the regression equation in (a) to compute the predicted value. Your answers should be very close.

(c) Which regression quantity measures the percent of the variation in the response explained by the straight-line relationship with the predictor variable? Use software or a calculator to compute this quantity.

Exercise 2.77

KEY CONCEPTS: Units of measurement for descriptive measures, effect of transformations on summary measures

(a) The mean and standard deviation are discussed in Section 2 of Chapter 1 of the text. The correlation is discussed in Section 2 of this chapter. Refer to these sections for help if you have forgotten how to compute the mean, standard deviation, or correlation, or if you have forgotten the units of measurement for these quantities. Compute $\bar{x}$, s_x, $\bar{y}$, s_y, and r and then enter the values below. These computations can be done easily using software.

$\bar{x}$ = ____________ s_x = ____________

$\bar{y}$ = ____________ s_y = ____________

r = ____________

What are the units of measurement for each of these descriptive measures? Be sure to include these in the values of your summary statistics.

(b) You need to determine which of these descriptive measures would have new units of measurement and new values and what effect the linear transformation inches = 2.54 × centimeters has on each quantity.

$\bar{y}$ = ____________ s_y = ____________

r = ____________

(c) Recall that the slope is $b = r\frac{s_y}{s_x}$. Compute this from the quantities you calculated in parts (a) and (b).

b = ____________

Exercise 2.79

KEY CONCEPTS: r versus r^2

Recall that r^2 tells us the fraction of the variation in y that is explained by the regression of y on x. Is the number 16% the value of r or r^2? If it is r^2, what is the sign of r?

COMPLETE SOLUTIONS

Exercise 2.63

(a) The slope measures the increase (on the average) of the volume of water discharged with a change in time of 1 year or, equivalently, with each passing year. The volume of water increases (on average) 4.2255 cubic kilometers with each passing year.

(b) At $x = 1780$, the predicted value of y is

$$\hat{y} = a + bx = -7792 + (4.2255 \times 1780) = -270.61$$

The timeplot covers only the years 1954 to 2001. The year 1780 is well outside the range of the data. Because the answer is negative, we know that the volume of water discharged couldn't have increased at this constant rate since 1780 since the answer is negative which does not make sense in this problem.

(c) At $x = 1990$, the predicted value of y is

$$\hat{y} = a + bx = -7792 + (4.2255 \times 1990) = 616.75$$

Using Table 1.4, we see that the observed value of y in 1990 is 680 cubic kilometers. Thus, we have

$$\text{Error} = \text{observed } y - \text{predicted } y = 680 - 616.75 = 63.25 \text{ cubic kilometers}$$

(d) A flood would cause the observed y to be larger than that predicted by the regression equation, as this is an unusual circumstance not accounted for by the general pattern of the data. In 1973 and 1993 the actual discharges were 880 and 900 cubic kilometers, respectively. Examination of Figure 1.10 shows large "spikes" at 1973 and 1993 corresponding to values of annual discharge much larger than predicted by the regression equation.

Exercise 2.67

(a, b) The plot on the next page gives the scatterplot with the least-squares regression line drawn on it. Many software packages will do these plots automatically as part of the regression output. The "up and over" lines show that the predicted brain activity for a social distress score of 2.0 is slightly less than zero. Although difficult to determine from the graph with much accuracy, we could say that it was between –0.01 and 0, but closer to zero. Using the regression equation in part (a) we have

```
Activity = -0.126 + 0.0608 Distress = -0.126 + (0.0608 × 2) = -0.0044.
```

This agrees fairly closely with the approximate predicted activity using the "up and over" method.

(c) The value of r^2 gives the percent of the variation in y (brain activity) that is explained by the straight-line regression with x (social distress score). Software gives $r^2 = 77.1\%$.

Exercise 2.77

(a) Using statistical software we found

$\bar{x} = 95$ $\quad s_x = 53.3854$

$\bar{y} = 12.6611$ $\quad s_y = 8.4967$

$r = 0.996$

The units of measurement are $\bar{x}$ = minutes, $\bar{y}$ = centimeters, s_x = minutes, s_y = centimeters, and r = unitless.

(b) The linear transformation inches = 2.54 × centimeters changes the mean from $\bar{y}$ in centimeters to (2.54 × $\bar{y}$ in inches. Likewise, it changes the standard deviation from s_y in centimeters to 2.54 × s_y in inches. The correlation r is unchanged because it is unitless. Thus, we get

new $\bar{y}$ in inches $= 12.6611 \times 2.54 = 32.1592$ inches

new s_y in inches $= 8.4967 \times 2.54 = 21.5816$ inches

new $r = 0.996$ inches/minute

(c) We compute

$$b = r\frac{s_y}{s_x} = 0.996 \times \frac{21.5816}{53.3854} = 0.4026$$

Exercise 2.79

16% is the value of r^2. Hence, $r^2 = 0.16$ and $r = \sqrt{0.16} = 0.4$. We take the positive square root because the problem states that, in general, students who attended a higher percent of their classes earned a higher grade, which corresponds to a positive association.

SECTION 2.4

OVERVIEW

Plots of the **residuals,** which are the differences between the observed and predicted values of the response variable, are very useful for examining the fit of a regression line. Features to look out for in a residual plot are unusually large values of the residuals (outliers), nonlinear patterns, and uneven variation about the horizontal line through zero (corresponding to uneven variation about the regression line).

The effects of **lurking variables,** variables other than the explanatory variable that may also affect the response, can often be seen by plotting the residuals versus such variables. Linear or nonlinear trends in such a plot are evidence of a lurking variable. If the time order of the observations is known, it is good practice to plot the residuals versus time order to see if time can be considered a lurking variable.

Influential observations are individual points whose removal would cause a substantial change in the regression line. Influential observations are often outliers in the horizontal direction but they need not have large residuals.

Correlation and regression must be interpreted with caution. Plots of the data, including residual plots, help to make sure the relationship is roughly linear and help to detect outliers and influential observations. The presence of lurking variables can make a correlation or regression misleading. *Always remember that association, even strong association, does not imply a cause-and-effect relationship between two variables.*

A correlation based on averages is usually higher than if we had data for individuals. A correlation based on data with a restricted range is often lower than would be the case if we had observed the full range of the variables.

GUIDED SOLUTIONS

Exercise 2.95

KEY CONCEPTS: Scatterplots, regression equation, influential observations

(a) Make your scatterplot on the axes provided on the next page. Except for one potentially influential observation, the remaining points follow a generally linear pattern with a moderate linear association. On your scatterplot, circle the influential observation.

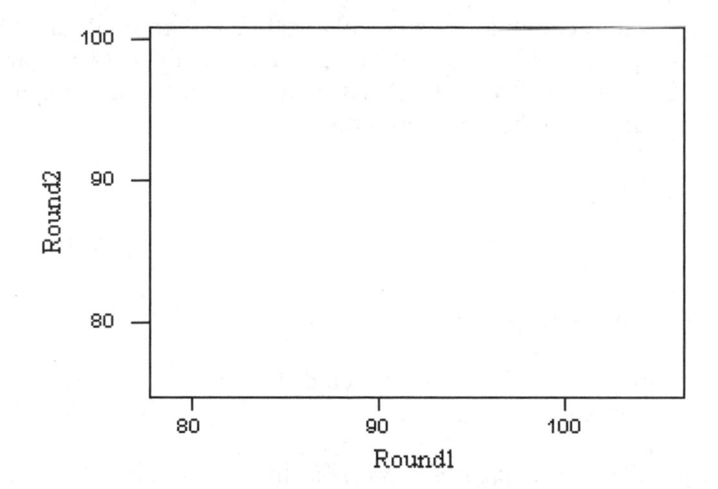

(b) Draw the two least-squares lines given on your scatterplot. Remember that for each line it is enough to find the predicted round2 score at two round1 scores, say 80 and 100, in order to draw the line on the graph. An influential observation may change the regression equation by "pulling" the regression line toward it. Can you tell which line is computed with the influential observation?

Exercise 2.98

KEY CONCEPTS: Regression equation, predicted values, residuals

(a) If you are using statistical software, you should enter the data and use the software to create a scatterplot. Although we are giving you many of the plots, it is a good idea to make sure you understand in each plot why one variable was designated the response and the other the explanatory variable.

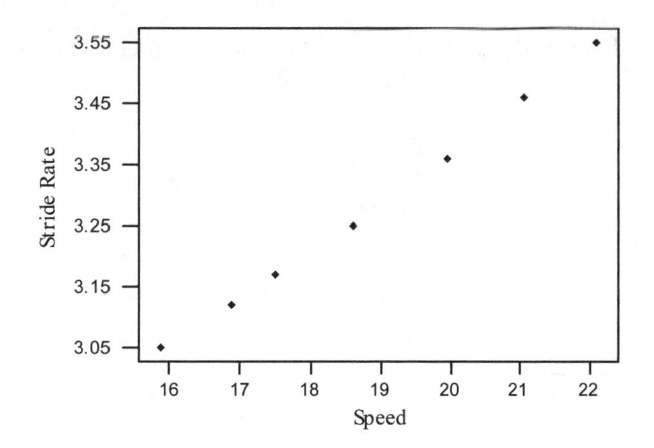

What is the general trend in your scatterplot? Does it appear to be adequately described by a straight line or is curvature present?

(b) If you do not have access to a computer or a calculator that will compute the least-squares regression line, you will have to do computations by hand. If you are using statistical software (or a calculator), you should enter the data and use the software (calculator) to calculate the equation of the least-squares regression line. Write the equation in the space provided below.

(c) Recall that the residual for a given speed x is

residual = observed stride rate – predicted stride rate
predicted stride rate = $a + bx$

and $a + bx$ is the equation of the least-squares regression line from part (b). Statistical software can be used to calculate the residuals directly. If you use statistical software, fill in the values in the residual column only in the table below. If you are calculating the residuals by hand, complete the table below to aid you in systematically calculating the residuals.

Speed	Observed stride rate	Predicted stride rate	Residual = observed – predicted
15.86	3.05		
16.88	3.12		
17.50	3.17		
18.62	3.25		
19.97	3.36		
21.06	3.46		
22.11	3.55		

Add the entries in the residual column to verify that the residuals sum to 0 (except for rounding error).

(d) If you are using statistical software, the software should allow you to create a plot of the residuals directly. Plot the seven residuals on the axes below.

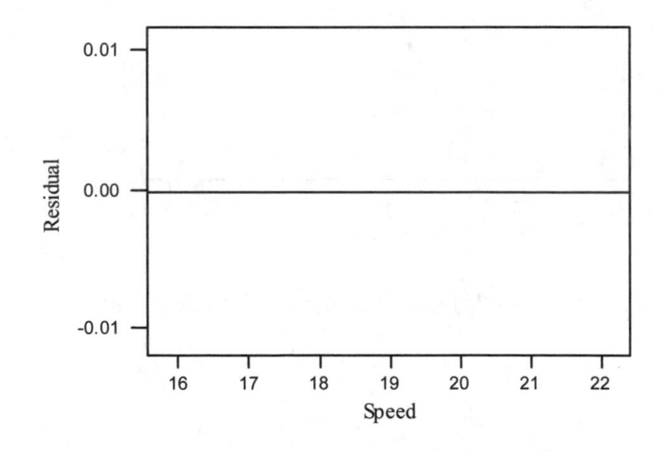

Does it appear that the residuals have a random scatter or is there a pattern present? Do you think that a linear fit is appropriate for these data?

Do any of the points in your plot appear to be influential? Ask yourself if their removal would cause a substantial change in the least-squares line (or if they appear to be outliers in the horizontal direction).

Note: For classes that discussed the topic of DFFITS, you can also calculate the DFFITS to determine if any observations are influential.

Are you provided with any information that would allow you to plot the observations against the time they were made? You might ask yourself, in what form would this information have to be in order to make such a plot? (Think about what the actual data values represent.)

Exercise 2.100

KEY CONCEPTS: Correlations based on averaged data

Computing the correlation is most easily done using statistical software or a calculator that computes correlation. Regarding whether the correlation would increase or decrease if we had data on the individual stride rates of all 21 runners, note that a correlation based on averages over many individuals is usually higher than the correlation between the same variables based on data for individuals.

COMPLETE SOLUTIONS

Exercise 2.95

(a, b) A scatterplot of the data with the influential observation circled is given below. The two regression lines, with and without this point, are drawn on the graph as well.

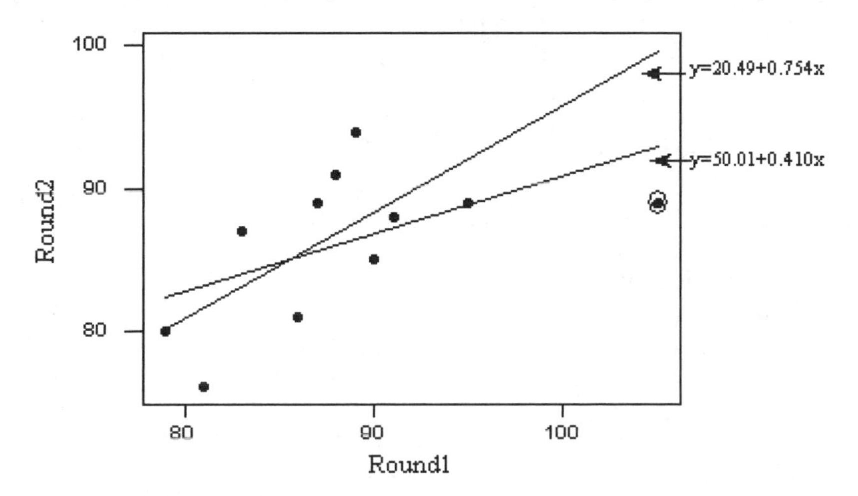

The line $y = 50.01 + 0.410x$ includes the outlier, and you can see that the circled point tends to pull the line towards it. The line $y = 20.49 + 0.754x$ fits better through the remaining 10 points.

Exercise 2.98

(a) A plot of the data is given in the Guided Solutions. A straight line appears to describe the data quite well.

(b) The least-squares regression line is

stride rate = 1.77 + 0.08(speed)

A scatterplot with this line is given below.

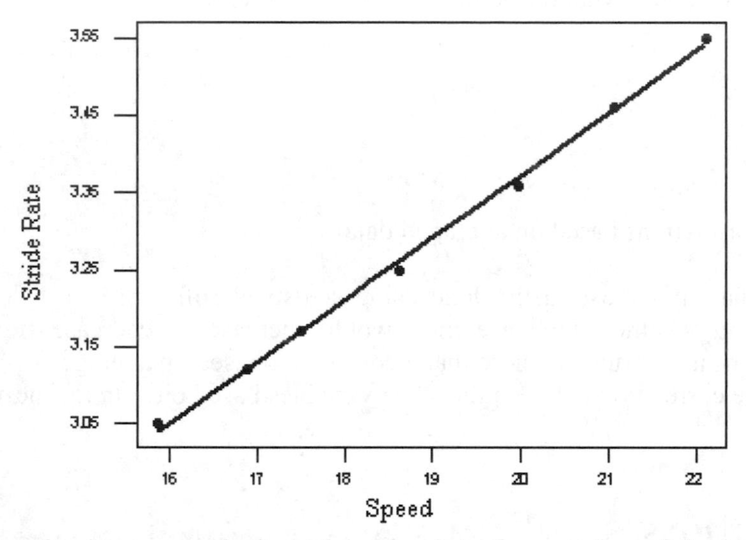

(c) For each of the speeds given, we substitute the value into the equation of the least-squares regression line to compute the predicted value of stride rate. The residual is then computed as

residual = observed stride rate – predicted stride rate

For example, for a speed of $x = 15.86$ we compute

predicted stride rate = 1.77 + 0.08(15.86) = 1.77 + 1.27 = 3.04

and hence

residual = observed stride rate – predicted stride rate = 3.05 - 3.04 = 0.01

We summarize the results in the table below (to two decimal places).

Speed	Observed stride rate	Predicted stride rate	Residual
15.86	3.05	1.77 + 0.08(15.86) = 3.04	0.01
16.88	3.12	1.77 + 0.08(16.88) = 3.12	0.00
17.50	3.17	1.77 + 0.08(17.50) = 3.17	0.00
18.62	3.25	1.77 + 0.08(18.62) = 3.26	–0.01
19.97	3.36	1.77 + 0.08(19.97) = 3.37	–0.01
21.06	3.46	1.77 + 0.08(21.06) = 3.46	0.00
22.11	3.55	1.77 + 0.08(22.11) = 3.54	0.01

While these calculations can be done by hand, they are much more easily obtained using statistical software and should agree (to two decimal places) with the above. If you sum the entries in the residual column, you can easily see they sum to 0.

(d) A plot of the residuals against speed is given below.

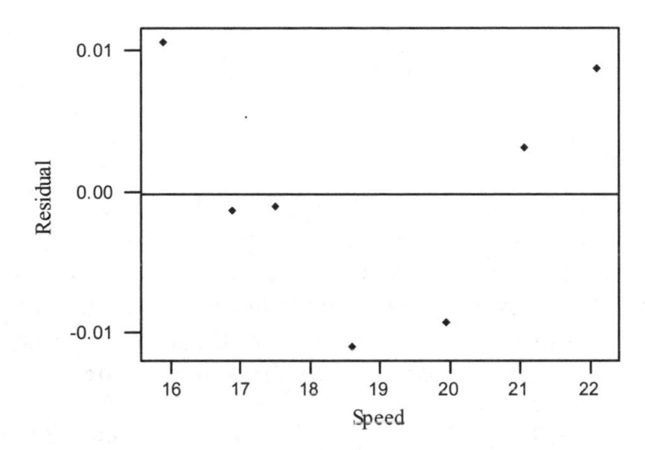

The pattern is curved (U-shaped) and indicates that the linear fit is not completely adequate. Recall that the scatterplot in (b) suggested that the linear fit was quite good. Reinspection of this plot shows the slight curvature, although it is much more apparent in the residual plot. This demonstrates that the residual plot can be more informative than the scatterplot concerning the adequacy of the linear fit.

There do not appear to be any influential observations in the plot. Since we are not told the time at which observations were made, we cannot plot the residuals against time. Note that since observations are actually the averages for 21 runners, we would have to know that all observations on the 21 runners at a given speed were made at the same time in order to even be able to make such a plot.

Note: For classes that discussed DFFITS, the values of DFFITS are given below. These are most easily computed using statistical software.

Speed	Stride rate	DFFITS
15.86	3.05	1.6635023
16.88	3.12	-0.08839246
17.50	3.17	-0.05792908
18.62	3.25	-0.59243516
19.97	3.36	-0.55913409
21.06	3.46	0.24486537
22.11	3.55	1.4562468

The first and last entries listed are the largest values of DFFITS and hence the most influential observations. Neither point appears to be particularly influential, however.

Exercise 2.100

The value of $r = 0.999$. This correlation is based on averaged data, namely the average stride rates of 21 runners at each of the seven values of speed (note that such data might have been collected by having the runners run on a treadmill where speed can be controlled). If we had the data on the individual stride rates of all 21 runners, we would expect the correlation to decrease (be less than 0.999).

SECTION 2.5

OVERVIEW

This section discusses techniques for describing the relationship between two or more categorical variables. To analyze categorical variables, we use counts (frequencies) or percents (relative frequencies) of individuals that fall into various categories. **A two-way table** of such counts is used to organize data about two categorical variables. Values of the **row variable** label the rows that run across the table, and values of the **column variable** label the columns that run down the table. In each cell (intersection of a row and column) of the table, we enter the number of cases for which the row and column variables have the values (categories) corresponding to that cell.

The **row totals** and **column totals** in a two-way table give the **marginal distributions** of the two variables separately. It is usually clearest to present these distributions as percents of the table total. Marginal distributions do not give any information about the relationship between the variables. **Bar graphs** are a useful way of presenting these marginal distributions.

The **conditional distributions** in a two-way table help us to see relationships between two categorical variables. To find the conditional distribution of the row variable for a specific value of the column variable, look only at that one column in the table. Express each entry in the column as a percent of the column total. There is a conditional distribution of the row variable for each column in the table. Comparing these conditional distributions is one way to describe the association between the row and column variables, particularly if the column variable is the explanatory variable. When the row variable is explanatory, find the conditional distribution of the column variable for each row and compare these distributions. Side-by-side bar graphs of the conditional distributions of the row or column variable can be used to compare these distributions and describe any association that may be present.

Data on three categorical variables can be presented in a **three-way table,** printed as separate two-way tables for each value of the third variable. An association between two variables that holds for each level of this third variable can be changed, or even reversed, when the data are **aggregated** by summing over all values of the third variable. **Simpson's paradox** refers to a reversal of an association by aggregation.

GUIDED SOLUTIONS

Exercise 2.111

KEY CONCEPTS: Joint and marginal distributions in two-way tables

For convenience, on the next page we reproduce the table given in the problem. The entries are in thousands of U.S. college students.

U.S. college students by age and status

Age	Status		
	Full-time	Part-time	Total
15 – 19	3388	389	
20 – 24	5238	1164	
25 – 34	1703	1699	
35 and over	762	2045	
Total			

(a) For this part you want the number of full-time college students aged 15 to 19. Which entry corresponds to this number?

(b) Before beginning this part, you should fill in the columns in the table in part (a) corresponding to the totals as you will need them for the remainder of this problem. The joint distribution is the collection of cell proportions for the two categorical variables. A cell proportion is the proportion that each cell count is of the total number of counts in the table (in this case, 16,388 thousands of persons). For example, the cell proportion corresponding to the age group "15 – 19" and "Full-time" status is 3388/16,388 = 0.207.

To answer the question, you need to compute each cell proportion and enter them in the table below. This will be the joint distribution. You can double-check your work by verifying that the sum of all the proportions is 1.

U.S. college students by age and status

Age	Status	
	Full-time	Part-time
15 – 19	0.207	
20 – 24		
25 – 34		
35 and over		

(c) The marginal distribution of age can be found from the totals that you have computed for each age group in the rightmost column of the table. Each value in this column must be divided by the total number of counts, namely 16,638, to give the proportion in each age group. Compute each of these proportions and enter the results in the table below.

Age group	Proportion
15 – 19	
20 – 24	
25 – 34	
35 and over	

Now display the results of this table graphically on the axes provided on the next page. The bar for 25 to 34 is provided.

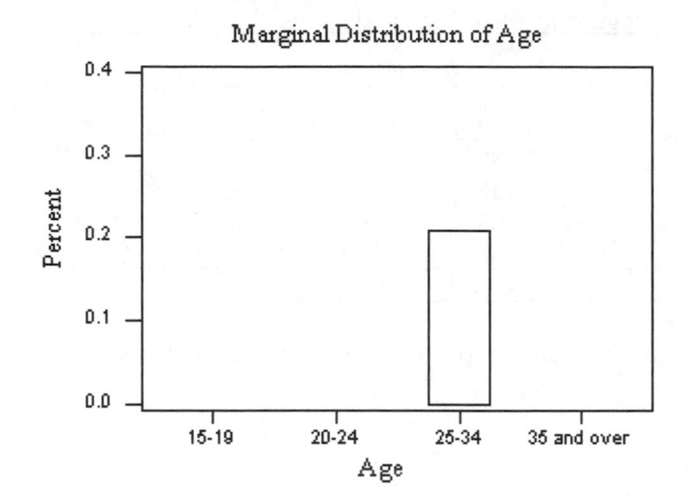

(d) The marginal distribution of status can be found from the totals for full-time and part-time status given in the bottom row of the table. Each value in this row must be divided by the total number of counts, namely 16,388, to give the proportion in each status. Compute each of these proportions and enter the results in the table below.

Status	Proportion
Full-time	
Part-time	

Now display the results of this table graphically on the axes provided.

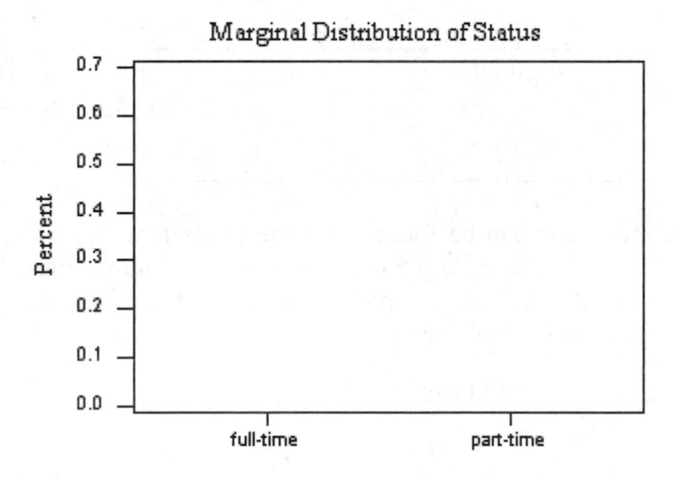

Exercise 2.113

KEY CONCEPTS: Conditional distributions and association in a two-way table

To compute the conditional distribution of age for a particular status category one must divide each cell entry in the row corresponding to the age by the row total (entry in bottom row of the table). For example, for the status "full-time" the conditional distribution involves dividing the entries 3388, 5238, 1703 and 762 each by 11,091.
Carry out these calculations and enter the results in the table below.

	Status	
Age	Full-time	Part-time
15 – 19		
20 – 24		
25 – 34		
35 and over		

To summarize these results graphically, make a separate bar graph for each status category. To assist you, we have provided the axes for the two plots below.

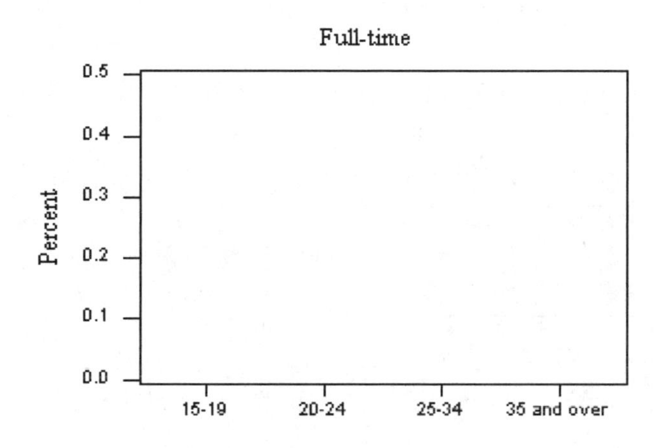

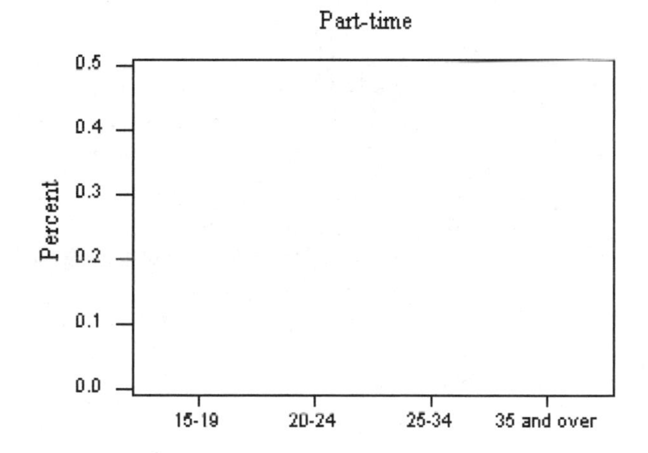

How do the distributions differ?

Exercise 2.159

KEY CONCEPTS: Aggregating three-way tables, Simpson's paradox

(a) Add corresponding entries in the three-way table and enter the sums in the table below.

	Admit	
Gender	Yes	No
Male		
Female		

(b) Convert your table in part (a) to one involving percents of the row totals.

	Admit	
Gender	Yes	No
Male		
Female		

(c) Now repeat the type of calculations you did in part (b) for each of the original tables.

Business

	Admit	
Gender	Yes	No
Male		
Female		

Law

	Admit	
Gender	Yes	No
Male		
Female		

(d) To explain the apparent contradiction observed in part (c), consider which professional school is easier to get into and which professional school males and females apply to in larger proportions. Write your answer in plain English in the space provided. Avoid jargon and be clear!

COMPLETE SOLUTIONS

Exercise 2.111

(a) The number of full-time college students aged 15 to 19 is the entry 3388 thousands of students.

(b) The joint distribution is as indicated in the completed table below.

U.S. College students by age and status

	Status	
Age	Full-time	Part-time
15 – 19	0.207	0.024
20 – 24	0.320	0.071
25 – 34	0.104	0.104
35 and over	0.046	0.125

(c) The marginal distribution of age and graphical display are given below.

	Age group			
	15 to 19	20 to 24	25 to 34	35 and over
Proportion	0.231	0.391	0.208	0.171

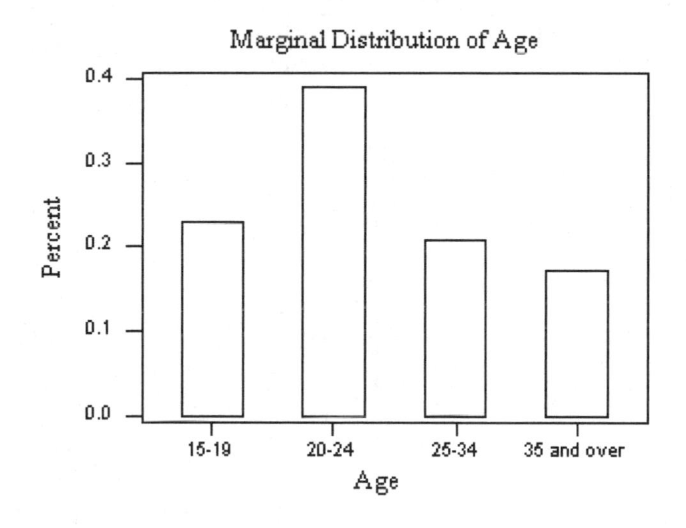

d) The marginal distribution of status and graphical display are given below.

Status	Proportion
Full-time	0.677
Part-time	0.323

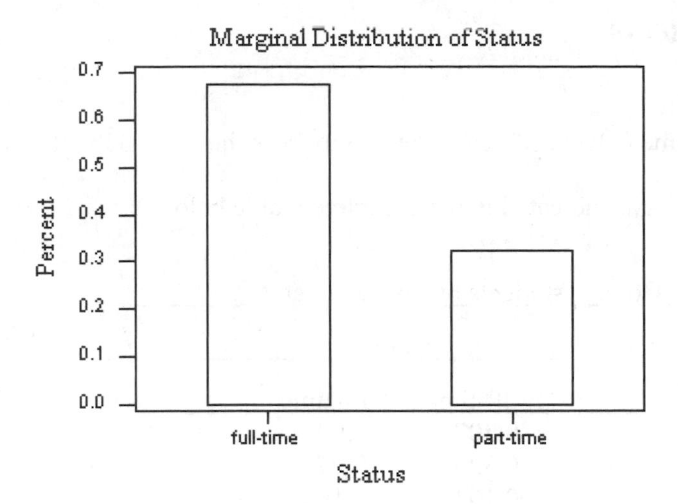

Exercise 2.113

The conditional distributions of age for full-time and part-time is summarized in the columns of the table below.

	Status	
Age	Full-time	Part-time
15 – 19	0.305	0.073
20 – 24	0.472	0.220
25 – 34	0.154	0.321
35 and over	0.068	0.386

Graphical displays of these conditional distributions follow.

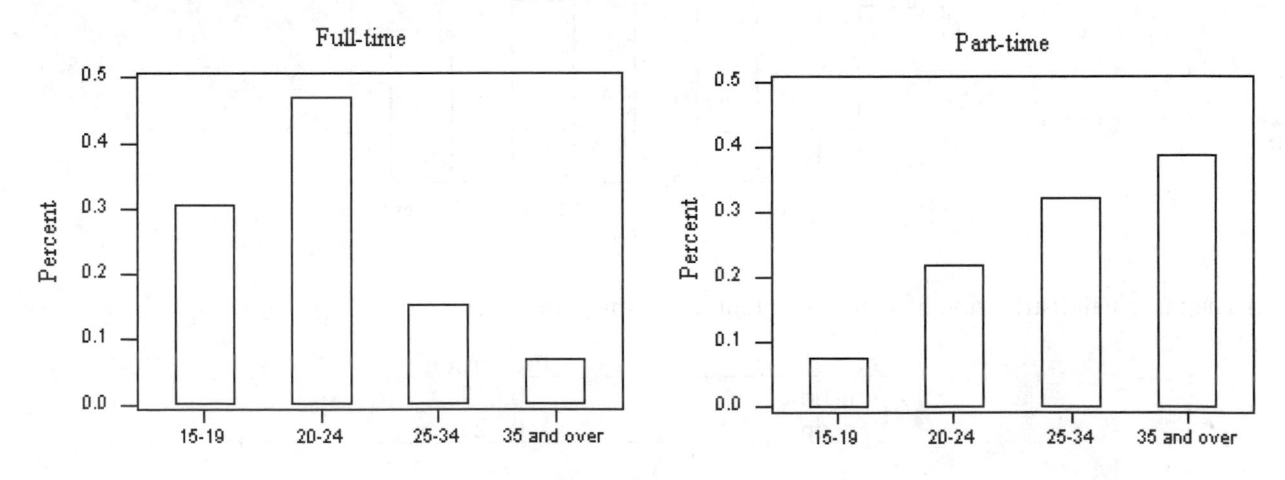

The age distributions of full-time and part-time students are quite different. The proportions in each age category for part-time increase with age, with the majority being in the 35 and over category. For the full-time students most are in the 15 to 19 and the 20 to 24 age categories. This is not surprising, as many part-time students are working and have families and are older than the full-time students, who have generally gone directly from high school to college.

Exercise 2.159

(a) To make the desired table we add together the entries in the corresponding cells of the two tables given. We get

Gender	Admit: Yes	No
Male	490	310
Female	400	300

(b) To compute the percents we must determine the row totals and then express each entry as a percent of the row total. The row totals are 800 for the Male row and 700 for the Female row. We then obtain the following.

Gender	Admit: Yes	No
Male	490/800 = 0.6125	310/800 = 0.3875
Female	400/700 = 0.5714	300/700 = 0.4286

We indeed see that a higher percent of males are admitted than females.

(c) We repeat the type of calculations in (b) but applied separately to the original tables given. Note that the row totals for Business are 600 males and 300 females, and for Law are 200 males and 400 females.

Business

Gender	Admit: Yes	No
Male	400/600 = 0.6667	200/600 = 0.3333
Female	200/300 = 0.6667	100/300 = 0.3333

Law

Gender	Admit: Yes	No
Male	90/200 = 0.4500	110/200 = 0.5500
Female	200/400 = 0.5000	200/400 = 0.5000

Now we see that each school admits at least as high a proportion of females as males.

(d) If we examine the data more closely, we see that Business tends to admit a higher percent of students than Law. We also see that the majority of applicants to Business are male, while the majority of applicants to Law are female. The fact that more males than females apply to Business (where it is easier to get admitted) while more females than males apply to Law (where it is harder to get admitted) means that the overall admission rate for females will appear relatively low (since they are applying to Law, which is hard to get into) compared to the admission rate for males (since males are applying to Business, which is easy to get into).

SECTION 2.6

OVERVIEW

An observed association between two variables can be due to several things. It can be due to a **cause-and-effect** relationship between the variables. It can also be due to the effects of **lurking variables**, i.e., variables not directly studied that may affect the response and possibly the explanatory variable. Lurking variables may operate through **common response,** in which case changes in both the explanatory and response variables are caused by changes in the lurking variable. Lurking variables may also cause **confounding,** in which case both the explanatory variable and the lurking variables cause changes in the response, but we cannot distinguish their individual effects.

The best way to determine if an association is due to a cause-and-effect relationship between the explanatory variable and the response variable is through an **experiment** in which we control the influences of other variables. In the absence of good experimental evidence, be cautious in accepting claims of causation. Good evidence requires an association that appears consistently in many studies, a clear explanation for the alleged cause-and-effect relationship, and careful examination of possible lurking variables.

GUIDED SOLUTIONS

Exercise 2.129

KEY CONCEPTS: Lurking variables, explaining association

The idea is that the association between chemicals and miscarriages may be explained by the lurking variable "time standing." The causal link may be that time standing results in the miscarriages, not the exposure to chemicals. Illustrate these relationships with a diagram like those in Figure 2.29.

Exercise 2.131

KEY CONCEPTS: Confounding variables

Ask yourself the following questions.

• Was the study an **experiment** in which the influences of other variables were controlled?

• If the study was not an experiment, what are some other variables that might be confounded with heavy TV viewing and possibly contribute to poor grades?

Exercise 2.132

KEY CONCEPTS: Explaining causation, lurking variables

Ask yourself, what type of people tend to use artificial sweeteners and why are they using them? This should help you give a more plausible explanation for this association.

COMPLETE SOLUTIONS

Exercise 2.129

The diagram below is like Figure 2.29(b). "Time standing" plays the role of z, "miscarriages" is the response y, and "exposure to chemicals" the explanatory variable x.

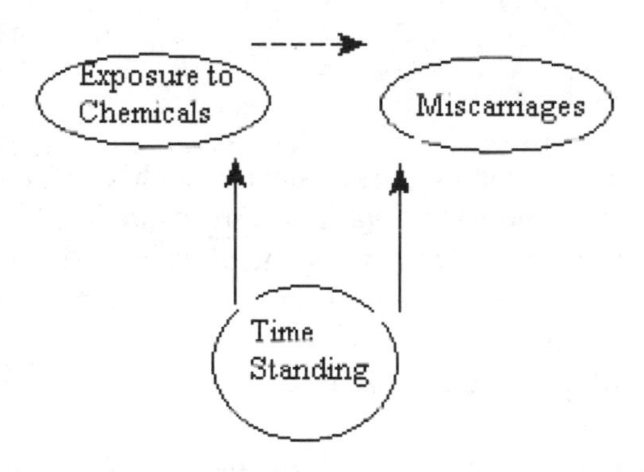

Exercise 2.131

Heavy TV viewing may be associated with confounding variables that are associated with poor grades. Many children from lower socioeconomic levels may spend a great deal of time watching TV, as it is the cheapest form of "entertainment" available. In some families, TV may serve as a substitute for parental involvement with children. In both of these instances, there may also be less emphasis on education, which could result in poorer grades.

Exercise 2.132

Many people begin to use artificial sweeteners to help control their weight. Those concerned about controlling their weight tend to be heavier than those who are not concerned about this. Diabetics also avoid sugar and as a group tend to be heavier.

CHAPTER 3

PRODUCING DATA

INTRODUCTION

OVERVIEW

Chapters 1 and 2 describe methods for exploring data. Such **exploratory data analysis** is used to determine what the data tell us about the variables measured and their relations to each other. Conclusions apply to the data observed and may not generalize beyond these data.

Statistical inference produces answers to specific questions, along with a statement of how confident we are that the answer is correct. Answers are usually intended to apply beyond the data observed. This requires careful **production of data** that are appropriate for answering the specific questions asked.

Data can be produced in many ways. **Anecdotal data** based on a few isolated cases are usually unreliable. **Available data** collected for other purposes, such as data produced by government agencies, can be helpful but again are not always reliable. **Sampling** selects a part of a population of interest to represent the whole. Done properly, sampling can yield reliable information about a population. **Observational studies** are investigations in which one simply observes the state of some population, usually with data collected by sampling. Even with proper sampling, data from observational studies are generally not appropriate for investigating cause-and-effect relations between variables. **Experiments** are investigations in which data are generated by active imposition of some treatment on the subjects of the experiment. Properly designed experiments are the best way to investigate cause-and-effect relations between variables.

GUIDED SOLUTIONS

Exercise 3.7

KEY CONCEPTS: Explanatory and response variables, experiments and observational studies

What are the researchers trying to demonstrate with this study? What groups are being compared and did the experiment deliberately impose membership in the groups on the subjects to observe their responses? What are the explanatory and response variables?

Explanatory variable: Response variable:

Is this an observational study or an experiment (circle one)? Why?

Exercise 3.9

KEY CONCEPTS: Explanatory and response variables, experiments

Remember, experiments are investigations in which data are generated by *active* imposition of some treatment on the subjects. Was that done in this example?

To identify the explanatory and response variables, think about what the experimenter is trying to demonstrate with this study and what is going to be measured.

COMPLETE SOLUTIONS

Exercise 3.7

The explanatory variable is whether or not the person made use of handheld cellular phones (we assume on a regular basis), and the response is whether or not the person contracted brain cancer. No attempt was made to decide which individuals were going to make use of cellular phones (the treatment), so this is an observational study, not an experiment.

Exercise 3.9

This is an experiment because a treatment is imposed on the students. The explanatory variable is the teaching method used (standard or computer assisted). The response variable is the increase in knowledge of cell biology as measured by the increase in test score.

SECTION 3.1

OVERVIEW

Experiments are studies in which one or more **treatments** are imposed on experimental **units** or **subjects**. A treatment is a combination of levels of the explanatory variables, called **factors**. The **design** of an experiment is a specification of the treatments to be used and the manner in which units or subjects are assigned to these treatments. The basic features of well-designed experiments are **control**, **randomization**, and **replication.**

Control is used to avoid confounding (mixing up) the effects of treatments with other influences such as lurking variables. One such lurking variable is the **placebo effect,** which is the response of a subject to the fact of receiving any treatment whatsoever. The simplest form of control is **comparative experimentation,** which involves comparisons between two or more treatments. One of these treatments may be a **placebo** (fake treatment), and those subjects receiving the placebo are referred to as a **control group**.

Randomization uses a well-defined chance mechanism to assign subjects to treatments. It is used to create treatment groups that are similar, except for chance variation, prior to application of treatments. Randomized, comparative experiments are used to prevent **bias**, or systematic favoritism of certain outcomes. **Tables of random digits** or computer programs that generate random numbers are well-defined chance mechanisms that are used to carry out randomization. In either case, numerical labels are assigned to experimental units, and random numbers from the table or computer software determine which labels (units) are assigned to which treatments.

Replication is the use of many units in an experiment to reduce the effect of any chance variation between treatment groups arising from randomization. Replication increases the sensitivity of an experiment to differences in treatments.

Additional control in an experiment can be achieved by forming experimental units into **blocks** that are similar in some way which is thought to affect the response. In a **block design**, units are first formed into blocks and then randomization is carried out separately in each block. **Matched pairs** are a simple form of blocking used to compare two treatments. In a matched pairs experiment either the same unit (the block) receives both treatments in a random order or very similar units are matched in pairs (the blocks). In the latter case, one member of the pair receives one of the treatments and the other member receives the remaining treatment. Members of a matched pair are assigned to treatments using randomization.

Good experiments require attention to details. **Double-blind** experiments are ones in which neither the subject nor the person measuring the response is aware of what treatment is being used. **Lack of realism** in an experiment can prevent us from generalizing the results.

GUIDED SOLUTIONS

Exercise 3.25

KEY CONCEPTS: Identifying experimental units or subjects, factors, treatments, and response variables

You need to read the description of the study carefully. To identify the experimental units, ask yourself, exactly on what were the experimental conditions applied?

To identify the factors, ask yourself, what question did the experiment wish to answer? What variables does the description say the answer depends on? These are the factors.

To identify the treatments, ask yourself, what combinations of values of the factors were actually used in the experiment? These are the treatments.

To identify the response variables, ask yourself, what was measured on the subjects after exposure to the treatments? This is the response variable.

Exercise 3.27

KEY CONCEPTS: Design of an experiment, randomization

To begin, identify the subjects, the factors, the treatments, and the response variable. Now outline your design. Be sure to specify the following.

• How many treatments are there? Hence, how many groups of subjects must you form?

• How will you assign subjects to treatment groups?

• What are the treatments, i.e., what will each subject be required to do?

• What response will you measure and how will you decide if the treatments differ in their effect?

You can outline your design in words or with a picture.

The list of names has been reproduced below. Assign a numerical label to each. Be sure to use the same number of digits for each label. So that everyone does the problem in the "same" way, we used the convention of starting with the label 00 for Anderson and continued to label across the row (Archberger 01, etc.), so that they are numbered in alphabetical order. Of course, one could start with another number (such as 01) and label down the columns if one wished.

Anderson	Archberger	Bezawada	Cetin	Cheng
Chronopoulou	Codrington	Daggy	Daye	Engelbrecht
Guha	Hatfield	Hua	Kim	Kumar
Leaf	Li	Lipka	Lu	Martin
Mehta	Mi	Nolan	Olbricht	Park
Paul	Rau	Saygin	Shu	Tang
Towers	Tyner	Vasilev	Wang	Watkins
Xu				

Now start reading line 151 in Table B. Read across the row in groups of digits equal to the number of digits you used for your labels (e.g., if you used two digits for labels, read line 151 in pairs of digits). You will need to keep reading until you have selected all the names for the first treatment. This may require you to continue on to line 152, line 153, and subsequent lines. After you have selected the names for Treatment 1, continue in Table B to assign the nine people to receive Treatment 2 and then nine to receive Treatment 3. The remaining names are assigned to Treatment 4.

Exercise 3.43

KEY CONCEPTS: Matched pairs design, randomization

The first thing you should do is identify the subjects, the treatments, and the response variable. Next, decide what are the matched pairs in this experiment. How will you use the table of random numbers to assign members of a pair to the treatments? Why is it important to have each player's trials on different days?

The first 20 digits of Table B at line 140 are reproduced below. Use these to decide which players will get oxygen on their first trial.

12975 13258 13048 45144

Exercise 3.46

KEY CONCEPTS: Properties of random digits

Write your answers (True or False) in the space provided. Remember that a table of random digits is defined to be a list of the digits 0, 1, 2, 3, 4, 5, 6, 7, 8, 9 that has the following properties:

1. The digit in any position in the list has the same chance of being any one of 0, 1, 2, 3, 4, 5, 6, 7, 8, 9.

2. The digits in different positions are independent in the sense that the value of one has no influence on the value of any other.

The text provides several consequences of these basic properties that may be helpful in completing the exercise.

(a)

(b)

(c)

Complete Solutions

Exercise 3.25

The experimental units are the adults contacted from the selected households.

There are two factors in this study, which are indicated by the introductory remarks. One factor is the information provided about the caller (name only, university being represented only, or name and university being represented). The other factor is whether survey results were offered (yes or no).

The treatments are combinations regarding the information about the caller and whether the survey results were offered. Thus, there are six treatments: (1) name only and survey results offered, (2) name only and survey results not offered, (3) university represented only and survey results offered, (4) university represented only and survey results not offered, (5) name and university provided and survey results offered, and (6) name and university provided and survey results not offered.

The response variable is whether or not the interview was completed.

Exercise 3.27

In this case, the subjects are the 36 headache sufferers who have agreed to participate in the study. The two factors are antidepressant (placebo or antidepressant given) and stress management training (given or not given). These form the four treatments, which we label as

Treatment 1: Antidepressant and no stress management training.

Treatment 2: Placebo (no antidepressant) and no stress management training.

Treatment 3: Placebo (no antidepressant) and stress management training.

Treatment 4: Antidepressant and stress management training.

The response variables are number of headaches over the study period and some measure of the severity of these headaches. The problem does not specify how the severity is to be measured.

The study should be done as follows. Subjects should be randomly assigned to treatments, with nine assigned to Treatment 1, nine assigned to Treatment 2, nine assigned to Treatment 3, and the remainder assigned to Treatment 4. Each subject follows his or her treatment regimen over the course of the study. The average number of headaches for each treatment should be calculated and the results for the four groups compared, as well as a comparison of the severity. A picture summarizing the experimental design is given below.

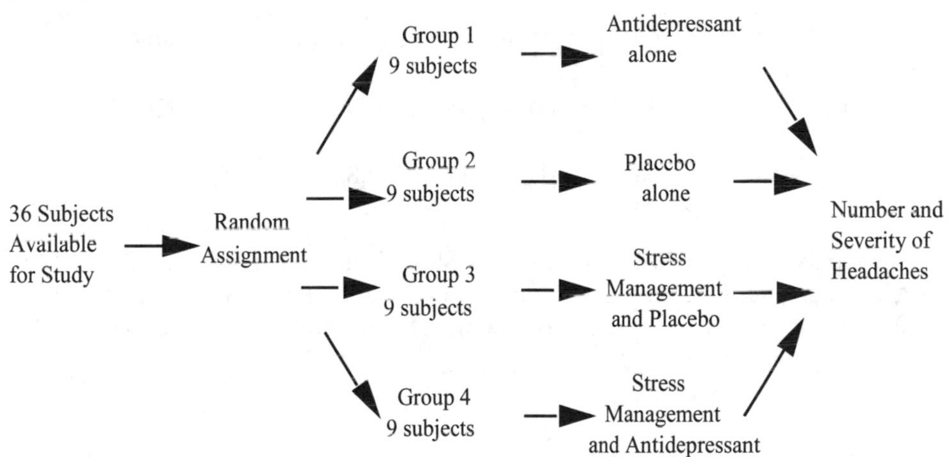

Although you can use the applet or Table B, we illustrate the use of Table B to carry out the random assignment of the subjects to the treatments. First label the 36 names using two-digit labels. We use the convention of starting with the label 00 and label across the rows. Of course, one could start with another number (such as 01) and label down the columns if one wished. The names with labels are

00 Anderson	01 Archberger	02 Bezawada	03 Cetin	04 Cheng
05 Chronopoulou	06 Codrington	07 Daggy	08 Daye	09 Engelbrecht
10 Guha	11 Hatfield	12 Hua	13 Kim	14 Kumar
15 Leaf	16 Li	17 Lipka	18 Lu	19 Martin
20 Mehta	21 Mi	22 Nolan	23 Olbricht	24 Park
25 Paul	26 Rau	27 Saygin	28 Shu	29 Tang
30 Towers	31 Tyner	32 Vasilev	33 Wang	34 Watkins
35 Xu				

Line 151 from Table B is reproduced below. We should read line 151 in pairs of digits from left to right. We have placed vertical bars between consecutive pairs to indicate how we have read the table. We underline those pairs that correspond to labels in our list and that have not been previously selected. On line 151 we have

03|80|2 2|93|41| 29|26|4 8|01|98| 12|37|1 1|31|21| 54|96|9 4|39|12

The first 9 labels give the Treatment 1 group. The Treatment 1 group consists of Cetin, Nolan, Tang, Rau, Archberger, Hua, Hatfield, Tyner, and Mi.

Continuing in line 151 and then going to line 152, line 153, and line 154 (making sure to ignore pairs corresponding to previously selected subjects), we assign the next nine subjects to Treatment 2,

| 54|96|9 4|39|12

77|32|0 3|50|30| 77|51|9 4|11|09| 98|29|6 1|89|84| 60|86|9 1|23|49

07|88|6 5|68|66| 39|64|8 6|92|90| 03|60|0 0|53|76| 58|95|8 2|27|20

87|06|

giving the Treatment 2 group as Vasilev, Towers, Engelbrecht, Olbricht, Daggy, Anderson, Saygin, Mehta, and Codrington. Continuing in line 154,

|5 7|41|33| 21|11|7 7|05|95| 22|79|1 6|73|06| 28|42|0 5|20|67

42|09|0 0|96|28| 54|03|5 9|38|79| 98|44|1 0|46|06| 27|38|1 8|26|37

55|49|4 6|76|90| 88|13|1 8|18|00| 11|18|8 2|85|52| 25|75|2 2|19|53

the Treatment 3 group is Wang, Chronopoulou, Li, Shu, Guha, Lu, Kim, Paul, and Martin. The nine remaining subjects, Bezawada, Cheng, Daye, Kumar, Leaf, Lipka, Park, Watkins, and Xu, are assigned to Treatment 4. Using Table B can become tedious with a large number of subjects and it is best to leave such calculations to a computer.

Exercise 3.43

We have 20 players available and there are two treatments. Treatment 1 is inhaling oxygen during the rest periods, and Treatment 2 is not inhaling oxygen between the rest periods. The response is the time on the final run.

To do the experiment, we use a matched pairs design. Each player is a block and each player will run 100 yards four times under each treatment, three in quick succession followed by a fourth time after a three-minute rest. We should randomly decide which treatment to use first. In order that the players have a chance to recover, we should allow sufficient time between the two trials (say, a day or two).

You are asked to use the table of random digits to assign which treatment comes first for each player. Suppose the players are numbered from 1 to 20. You will go to Table B, line 140, to decide which players get oxygen on their first trial. We will use the following randomization. If the random digit is 0, 1, 2, 3, or 4, then the subject gets oxygen on the first trial. Otherwise, they get oxygen on the second trial. The first 20 digits in line 140 are reproduced below, with the digits 0, 1, 2, 3, and 4 underlined.

12975 13258 13048 45144

Since positions 1, 2, 6, 7, 8, 11, 12, 13, 14, 16, 18, 19, and 20 correspond to digits 0, 1, 2, 3, or 4, players numbered 1, 2, 6, 7, 8, 11, 12, 13, 14, 16, 18, 19, and 20 receive oxygen during the rest periods on their first trial and no oxygen during their second trial. The remaining players get oxygen during the rest periods on their second trial and none during their first trial. In this example we happen to have 13 players get oxygen during the rest periods on their first trial and 7 get oxygen during the rest period on their second trial. When using a table of random numbers or software to generate random numbers, there is no guarantee that the groups will balance out exactly every time the randomization is done.

As an alternative, you could label the players 01 through 20. Begin on line 140 and select 10 of these at random to be in the oxygen group. The remaining 10 would not get oxygen during the rest period. This would guarantee balance.

Exercise 3.46

(a) This is false. Randomness does not mean each digit appears the exact same number of times in each row. For example, look at line 150 in Table B. There are only two 0s in this row.

(b) True. Randomness means each digit has an equal chance (1/10) of being a 0, each pair an equal chance (1/100) of being 00, each triple an equal chance (1/1000) of being 000, etc. This point is made in the section "How to Randomize" in the text a few paragraphs below the box containing the definition of a table of random digits.

(c) False. Following the logic in (b), any set of four digits has an equal chance (1/10,000) of being 0000. Thus, the digits 0000 can appear, but the chance of any four digits being this sequence is quite small.

SECTION 3.3

OVERVIEW

The **population** is the entire group of individuals or objects about which we want information. The information collected is contained in a **sample,** which is the part of the population we actually get to observe. How the sample is chosen, that is, the **design,** has a large impact on the usefulness of the data. A useful sample will be representative of the population and will help answer our questions. "Good" methods of collecting a sample include the following:

probability samples
simple random samples, also called **SRS**
stratified random samples
multistage samples

All these sampling methods involve some aspect of randomness through the use of a formal chance mechanism. Random selection is just one precaution that a person can take to reduce **bias,** the systematic favoring of a certain outcome. Samples we select using our own judgment, because they are convenient, or "without forethought" (mistaking this for randomness) are usually biased in some way. This is why we use computers or a tool like a **table of random digits** to help us select a sample.

A **voluntary response sample** includes people who choose to be in the sample by responding to a general appeal. They tend to be biased, as the sample is overrepresented by individuals with strong opinions, which are often negative.

Other kinds of bias to be on the lookout for include

nonresponse bias, which occurs when individuals who are selected do not participate or cannot be contacted,

undercoverage, which occurs when some group in the population is given either no chance or a much smaller chance than other groups to be in the sample, and

response bias, which occurs when individuals do participate but are not responding truthfully or accurately due to the way the question is worded, the presence of an observer, fear of a negative reaction from the interviewer, or any other such source.

These types of bias can occur even in a randomly chosen sample and we need to try to reduce their impact as much as possible.

GUIDED SOLUTIONS

Exercise 3.55

KEY CONCEPTS: Populations and sources of bias

(a) Try to identify the population as exactly as possible from the information given. What is the sample size?

(b) Describe the advantages and disadvantages of the two different ways of reporting the percent of favorable rating. Suppose a minor news personality gave fairly biased news coverage and had a small but loyal following. How would they rate on the two methods of reporting?

Exercise 3.57

KEY CONCEPTS: Selecting an SRS with a table of random numbers

The table of random numbers can be used to select an SRS of numbers. In order to use it to sample from the apartment complexes in the college town, the apartment complexes need to be assigned numbers. So that everyone does the problem the "same" way, we have numbered the apartment complexes according to alphabetical order:

01 – Ashley Oaks
02 – Bay Pointe
03 – Beau Jardin
04 – Bluffs
05 – Brandon Place
06 – Briarwood
07 – Brownstone
08 – Burberry
09 – Cambridge
10 – Chauncey Village
11 – Country Squire
12 – Country View
13 – Country Villa
14 – Crestview
15 – Del-Lynn
16 – Fairington
17 – Fairway Knolls
18 – Fowler
19 – Franklin Park
20 – Georgetown
21 – Greenacres
22 – Lahr House
23 – Mayfair Village
24 – Nobb Hill
25 – Pemberly Courts
26 – Peppermill
27 – Pheasant Run
28 – Richfield
29 – Sagamore Ridge
30 – Salem Courthouse
31 – Village Manor
32 – Waterford Court
33 – Williamsburg

If you go to line 137 in Table B and start selecting two-digit numbers until you have selected an SRS of 5 apartment complexes, you should get the same answer as given in the complete solution.

Exercise 3.64

KEY CONCEPTS: Systematic sampling

(a) This is like the example in the exercise, except there are now 9000 male students instead of 100 addresses and the sample size is now 200 instead of 4. With these two changes, you need to think about how many different systematic samples there are. The labels of individuals in the different systematic samples are

systematic sample 1 = 01, 46, 91, 136, 181, ⋯ , 8956
systematic sample 2 = 02, 47, 92, 137, 182, ⋯ , 8957

....

systematic sample 45 = 45, 90, 135, 180, 182, ⋯ , 9000

Why are you choosing every 45th name?

(b) How many systematic samples are there altogether? Choosing one of these systematic samples at random is equivalent to choosing the first person in the sample. The remaining 199 individuals follow automatically by adding 45 repeatedly as in the examples above. Carry this out using line 125 in the table. The first 10 digits are reproduced below. Use these to choose the starting point for the systematic sample.

96746 12149

Exercise 3.65

KEY CONCEPTS: Systematic sampling

Why are all individuals equally likely to be selected? First, how many systematic samples contain each individual? The chance of selecting an individual is the same as the chance of selecting the systematic sample that contains him. With this in mind, what is the chance of any individual being chosen? By the definition of an SRS, all samples of 200 individuals are equally likely to be selected. In a systematic sample, are all samples of 200 individuals even possible?

Exercise 3.70

KEY CONCEPTS: Stratified random sample

What are the two strata from which you are going to sample? A stratified random sample consists of taking an SRS from each stratum and combining these to form the full sample. How large an SRS will be taken from each of the two strata? How would you label the units in each of the two strata from which you will sample? Fill in the following table to describe your sampling plan.

	STRATA	
	Midsize accounts	Small accounts
Number of units in stratum		
Sample size		
Labeling method		

In practice, you would probably use the table of random numbers to first select the SRS from the midsize accounts and then you would select the SRS from the small accounts. In this problem you are not going to select the full samples, but only the first five units from each stratum. Start in Table B at line 115 and first select 5 midsize accounts and then continue in the table to select 5 small accounts. Write the numerical labels below.

First 5 midsize accounts ______________________________________

First 5 small accounts ____

Exercise 3.71

KEY CONCEPTS: Nonresponse, response bias

In which of the two periods do you expect the nonresponse to be higher? Is a particular group going to be underrepresented in one of the samples and could this make the sample less reliable?

Exercise 3.72

KEY CONCEPTS: Sampling frame, undercoverage

(a) Which households wouldn't be in the sampling frame? Make some educated guesses as to how these households might differ from those in the sampling frame (other than the fact that they don't have a phone number in the directory).

(b) Random digit dialing makes the sampling frame larger. Which households are added to it?

Exercise 3.74

KEY CONCEPTS: Wording of questions

A question can be worded in such a way to make it seem as though any reasonable person should answer Yes (or No). Which questions are slanted toward a desired response? Are all questions clear?

COMPLETE SOLUTIONS

Exercise 3.55

(a) The population of interest is adults aged 18 or older in the United States. A possible source of bias is that only residents with phones could be contacted, and if the phone numbers were selected from phone books, then residents with unlisted numbers could not be in the sample. The sample size is the 1001 adults who agreed to be interviewed.

(b) Reporting just a favorable or unfavorable opinion tends to make less familiar news personalities receive a smaller percentage with a favorable rating as many people will not have an opinion about them. This tends to favor popular news personalities on the major networks. A less familiar news personality such as Charles Gibson would do better when only those providing an opinion were counted. When including only those with an opinion, it is possible that a news personality would have a high approval rating among the selected group that are familiar with them but could have a highly unfavorable rating if more people were exposed to them.

Exercise 3.57

To choose an SRS of 5 apartment complexes for in-depth interviews, first label the members of the population by associating a two-digit number with each.

01 – Ashley Oaks
02 – Bay Pointe
03 – Beau Jardin
04 – Bluffs
05 – Brandon Place
06 – Briarwood
07 – Brownstone
08 – Burberry
09 – Cambridge
10 – Chauncey Village
11 – Country Squire
12 – Country View
13 – Country Villa
14 – Crestview
15 – Del-Lynn
16 – Fairington
17 – Fairway Knolls
18 – Fowler
19 – Franklin Park
20 – Georgetown
21 – Greenacres
22 – Lahr House
23 – Mayfair Village
24 – Nobb Hill
25 – Pemberly Courts
26 – Peppermill
27 – Pheasant Run
28 – Richfield
29 – Sagamore Ridge
30 – Salem Courthouse
31 – Village Manor
32 – Waterford Court
33 – Williamsburg

Now enter Table B and read two-digit groups until 5 apartment complexes are chosen. When a two-digit group appears a second time, skip it and go to the next two-digit group. Start at line 137:

53645 66812 61421 47836 12609 15373 98481 14592

66831 68908 40772 21558 47781 33586 79177 06928

The selected sample is 12=Country View, 14=Crestview, 11=Country Squire, 16=Fairington, and 08= Burberry.

Exercise 3.64

(a) We want to select 200 names out of the 9000. Because 9000/200 = 45, we can think of the list as 200 lists of 45 addresses. We choose one name from the first 45 at random, and then every 45th labeled name after that.

(b) The first step is to go to Table B, line 125, and choose the first two-digit random number you encounter that is one of the numbers 01, $\cdots$, 45.

96746 12149

The selected number is 21, so the sample includes male students with numbers 21, 66, 111, 156, 201, ⋯ , 8976.

Exercise 3.65

Each individual is in exactly one systematic sample, and the systematic samples are equally likely to be chosen. In the previous exercise, there were 45 systematic samples, each containing 200 individuals. The chance of selecting any individual is the chance of picking the systematic sample that contains him, which is 1 in 45.

A simple random sample of size n would allow every set of n individuals an equal chance of being selected. Thus, in this exercise, when using an SRS the sample consisting of the individuals numbered 1, 2, 3, 4, 5, ⋯ , 200 would have the same probability of being selected as any other set of 200 individuals. For a systematically selected sample, all samples of size n do not have the same probability of being selected. In our exercise the sample consisting of the individuals numbered 1, 2, 3, 4, 5, ⋯ , 200 would have zero chance of being selected since the numbers of the individuals do not all differ by 45. The sample we selected in Exercise 3.64 was 21, 66, 111, 156, 201, ⋯ , 8976, and had a 1 in 45 chance of being selected, so all samples of 200 individuals are not equally likely.

Exercise 3.70

There are 500 midsize accounts. We are going to sample 5% of these, which is 25. You should label the accounts 001, 002, ⋯ , 500 and select an SRS of 25 of the midsize accounts. There are 4400 small accounts. We are going to sample 1% of these, which is 44. You should label the accounts 0001, 0002, ⋯ , 4400 and select an SRS of 44 of the small accounts.

Starting at line 115, we first select 5 midsize accounts, that is, an SRS of size 5 using the labels 001 through 500. Continuing in the table we select 5 small accounts, that is, an SRS of size 5 using the labels 0001 through 4400. Note that for the midsize accounts we read from Table B using three-digit numbers, and for the small accounts we read from the table using four-digit numbers.

610|41 7|768|4 94|322| 247|09 7|3698| 1452|6 318|93 32|592 1|4459| 2605|6 314|24
80|371 6|

The first 5 midsize accounts are those with labels 417, 494, 322, 247, and 097. Continuing in the table, and using four digits instead of three, the first 5 small accounts are those with labels 3698, 1452, 2605, 2480, and 3716.

Exercise 3.71

You would expect that the higher rate of no-answer was probably during the second period as more families are likely to be gone for vacation. Nonresponse can always bias the results. In this case, those who are more affluent may be more likely to travel during the summer months and they would be underrepresented in the sample. Their views could be different, which would bias the results.

Exercise 3.72

(a) Households omitted from the frame are those that do not have a telephone number listed in the telephone directory. The types of people who might be underrepresented are poorer (including homeless) people who cannot afford to have a phone and people who have unlisted numbers. It is harder to characterize this second group. As a group they would tend to have more money because you need to pay to have your phone number unlisted, or it might include more single women who do not want their phone numbers available and possibly people whose jobs put them in contact with large groups of people who might harass them if their phone numbers were easily accessible.

(b) People with unlisted numbers will be included in the sampling frame. The sampling frame will now include any household with a phone. One interesting point is that all households will not have the same probability of getting in the sample, as some households have multiple phone lines and will be more likely to get in the sample. So, strictly speaking, random digit dialing will not actually provide an SRS of households with phones but just an SRS of phone numbers!

Exercise 3.74

(a) The beginning of the question suggests that cell phone use is associated with brain cancer. This initial suggestion and the wording "the danger of using cell phones" would lead most reasonable people to be in favor of including a warning label. The question is slanted in favor of this response.

(b) The question is clear but is slanted in favor of national health insurance. The reason for agreeing with a question should not be contained within the question.

(c) The question is slanted because it contains reasons why you should support recycling. As a question, the wording is a little technical for the general population, and a simpler version such as "Do you favor economic incentives to promote recycling?" would be better.

SECTION 3.3

OVERVIEW

Statistical inference is the technique that allows us to use the information in a sample to draw conclusions about the population. To understand the idea of statistical inference, it is important to understand the distinction between **parameters** and **statistics.** A **statistic** is a number we calculate based on a sample from the population—its value can be computed once we have taken the sample, but its value varies from sample to sample. A statistic is generally used to estimate a population **parameter,** which is a fixed but unknown number that describes the population.

The variation in a statistic from sample to sample is called **sampling variability**. It can be described through the **sampling distribution** of the statistic, which is the distribution of values taken by the statistic in all possible samples of the same size from the population. The sampling distribution can be described in the same way as the distributions we encountered in Chapter 1. Three important features are

- a measure of center
- a measure of spread
- a description of the shape of the distribution

The properties and usefulness of a statistic can be determined by considering its sampling distribution. If the sampling distribution of a statistic is centered (has its mean) at the value of the population parameter, then the statistic is **unbiased** for this parameter. This means that the statistic tends to neither overestimate nor underestimate the parameter.

Another important feature of the sampling distribution is its spread. If the statistic is unbiased and the sampling distribution has little spread or variability, then the statistic will tend to be close to the parameter it is estimating for most samples. The variability of a statistic is related to both the sampling design and the sample size n. Larger sample sizes give smaller spread (better estimates) for any sampling design. An important feature of the spread is that as long as the population is much larger than the sample (at least 100 times), the spread of the sampling distribution will depend primarily on the sample size, not the population size.

If the parameter p is the proportion of the population with a particular characteristic, then the statistic $\hat{p}$, the proportion in the sample with this characteristic, is an unbiased estimator. Provided the samples are selected at random, **probability** theory can be used to tell us about the distribution of a statistic.

GUIDED SOLUTIONS

Exercise 3.79

KEY CONCEPTS: Populations and samples

In identifying the population and sample you need to remember the population is the full set of individuals of interest and the sample is the individuals that we know about. Describe the population and sample for this setting.

Exercise 3.87

KEY CONCEPTS: Variability of the sample proportion

(a) "The variability of a statistic from a random sample does not depend on the size of the population, as long as the population is at least 100 times larger than the sample." You need to think about how this rule applies to this example.

(b) Is the rule given in part (a) applicable here? Read it carefully.

Exercise 3.91

KEY CONCEPTS: Sampling distributions

(a) The table of random numbers contains the 10 digits, 0, 1, 2, $\cdots$, 9, which are "equally likely" to occur in any position selected at random from the table. Selecting four digits from the table of random numbers will give an SRS of size 4 from the population of 10 students. For our sample we started at line 178 of the table, which is partially reproduced below.

24005 52114

The students selected have numbers 2, 4, 0, and 5 and scores 80, 72, 82, and 73. The mean $\bar{x}$ of the scores in this sample is 76.75.

b) For this part of the problem, everyone will be taking their 10 samples from different parts of the random number table. Some of you may know how to get random samples from your computer software. Work with your own samples here. Your answers will not agree exactly with that given in the complete solution but the general pattern should be similar. If everyone took 2000 samples instead of 10, would the sampling distributions from person to person show more or less agreement? List your 10 values of $\overline{x}$, including the one you obtained in (a), and draw your histogram below. Is the center close to 69.4?

COMPLETE SOLUTIONS

Exercise 3.79

The population consists of all undergraduate college students aged 18 to 24. The sample is the 2036 undergraduate college students who were included in the survey.

Exercise 3.87

(a) The population is at least 100 times the sample size $n = 2000$ for each of the states. So the variability in the sample proportion based on $n = 2000$ will be approximately the same for the population size of any state.

(b) The problem switches here. The rule applies to a given sample size: the variability of the sample proportion based on a fixed number of observations is approximately the same for any population size. Now the sample size will vary from state to state. For Wyoming, 1/10 of 1% of the population is a sample size of about $n = 494$, and 1/10 of 1% of the population of California is a sample size of about $n = 34{,}000$. Since larger sample sizes give smaller spread, there will be differences in the variability of the sample proportion from state to state. California's sample proportion will be much less variable than the sample proportion from Wyoming.

Exercise 3.91

(a) Done in the Guided Solutions.

(b) These are the values of $\overline{x}$ in the 10 samples we obtained using the computer to generate random digits, followed by the histogram.

Sample	$\overline{x}$
1	76.75
2	70.50
3	64.00
4	69.25
5	71.50
6	63.50
7	68.50
8	68.00
9	59.00
10	71.00

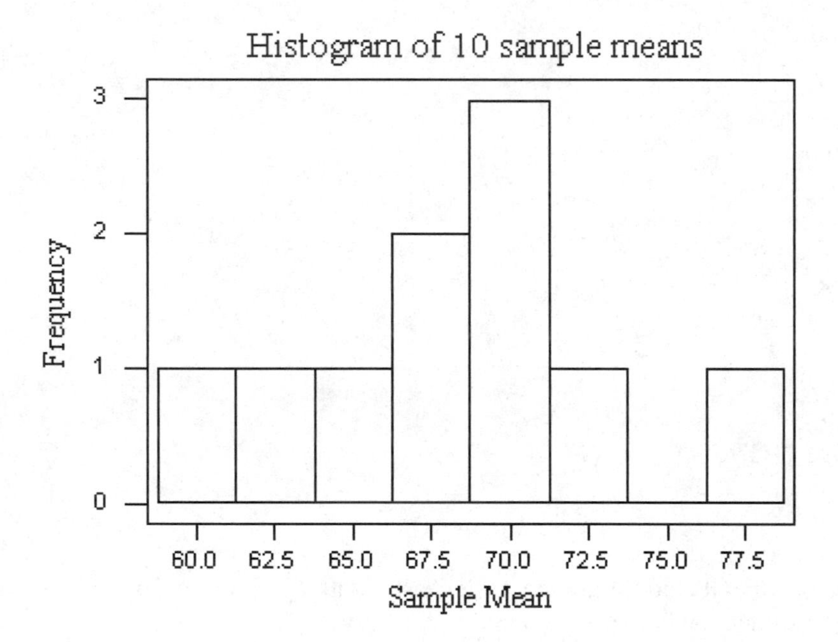

The mean of the distribution is 69.4. The histogram looks like it has a center fairly close to this. Your histogram may look quite different from this, but with only 10 samples, the distributions may vary quite a bit from person to person. If everyone took 2000 samples, which would require the sampling to be done using a computer, then the shapes of the distributions would be quite similar from person to person and the centers would be very close to 69.4.

SECTION 3.4

OVERVIEW

When conducting a study involving humans or animals as subjects, the study must have the approval of an **institutional review board**. When human subjects are involved, the subject must give **informed consent** to participate in a study. In addition, data on human subjects must be kept **confidential**.

GUIDED SOLUTIONS

Exercise 3.99

KEY CONCEPTS: Informed consent and confidentiality

Subjects must be informed in advance about the nature and purpose of the study and possible risks. In addition, the subject's privacy must be protected by keeping data about individuals confidential. In each of the following circumstances, would you allow collecting personal information without the subject's consent?

(a)

(b)

(c)

Exercise 3.105

KEY CONCEPTS: Confidentiality and anonymity

Confidentiality protects a subject's privacy by keeping all data about individuals confidential, while anonymity means that the names of the subjects are not known, even to the director of the study. From the description of the General Social Survey, are a subject's responses to the questions anonymous, confidential, or both? Explain.

Complete Solutions

Exercise 3.99

(a) Since the names of the subjects are not being recorded, the subject's consent should not be needed.

(b) If the members of the religious group object to being observed, then they should not be having public meetings which anyone can attend. The subject's consent should not be needed.

(c) This example is similar to those given in Example 3.40 of the text. As the social psychologist is attending private meetings under the false pretense that they have been converted to membership in the religious group, the informed consent of the subjects should be required. The subjects could have different behavior if they realized they were being observed. In psychology, this would violate the "Ethical Principles" of the American Psychological Association.

Exercise 3.105

The subject's responses are not anonymous as the interview is conducted in person in the subject's home. It would be difficult for the interviewer to get to a subject's home without knowing who they were. On the other hand, since the survey is designed to monitor public opinion, it is still possible to keep all data about the individual confidential. We would assume in this type of large scale governmental survey that confidentiality is protected.

CHAPTER 4

PROBABILITY: THE STUDY OF RANDOMNESS

SECTION 4.1

OVERVIEW

A process or phenomenon is called **random** if its outcome is uncertain. Although individual outcomes are uncertain, when the process is repeated many times the underlying distribution for the possible outcomes begins to emerge. For any outcome, its **probability** is the proportion of times, or the relative frequency, with which the outcome would occur in a long series of repetitions of the process. It is important that these repetitions or trials be **independent** for this property to hold.

You can study random behavior by carrying out physical experiments such as coin tossing or rolling a die, or you can simulate a random phenomenon on the computer. Using the computer is particularly helpful when we want to consider a large number of trials.

GUIDED SOLUTIONS

Exercise 4.1

KEY CONCEPTS: Random phenomena, probability

Since the probability of a head is 0.50, it is easy to use the table of random numbers to simulate a sequence of coin tosses. If 5 digits correspond to heads and the other 5 digits correspond to tails, then this makes heads and tails each have probability 0.50. In this case, we let the digits 0, 1, 2, 3, or 4 correspond to a head; otherwise a tail occurs. The first 20 digits from line 109 of Table B are reproduced below.

36009 19365 15412 39638

Since the first digit is a 3, the first toss of the coin is a head. Using this sequence to simulate 20 tosses,

Proportion of heads = ________

Why don't you get exactly 10 heads in 20 tosses?

Exercise 4.7

KEY CONCEPTS: Simulating a random phenomenon

(a) You can use your computer software or the applet to simulate the 100 trials, or you can use Table B. After simulating the 100 trials calculate the proportion of "hits."

proportion of hits = ________

For most students, their proportion of hits will be within 0.05 or 0.10 of the true probability of 0.5.

(b) You need to go through your sequence to determine the longest string of hits or misses.

Longest run of shots hit = ________ Longest run of shots missed = ________

COMPLETE SOLUTIONS

Exercise 4.1

Let the digits 0, 1, 2, 3, or 4 correspond to a head; otherwise a tail occurs. Using the first 20 digits form line 109 of Table B gives the sequence

```
36009   19365   15412   39638
HTHHT   HTHTT   HTHHH   HTTHT
```

and the proportion of heads = 11/20 = 0.55. Although the proportion of heads settles down to 0.50 in the long run, this doesn't mean that we would get exactly 10 out of 20 heads in 20 tosses. We expect the number of heads should be close to 10 (we will see how close in Chapter 5), but not exactly 10.

Exercise 4.7

(a) Our sequence of hits (H) and misses (M) is given below.

```
H  H  M  H  H  H  M  M  H  H  H  M  H  M  H
M  H  M  M  H  M  M  H  H  H  M  M  H  H  H
M  M  M  M  H  M  M  H  H  H  H  M  H  H  H
M  M  M  M  M  H  M  H  H  M  H  M  H  M  M
H  H  H  H  M  H  M  M  M  M  H  H  M  H  H
M  M  H  H  H  M  M  H  H  M  M  H  M  H  M
M  H  M  H  H  M  H  H  H  H
```

Proportion of hits = 0.54

(b) You need to go through your sequence to determine the longest string of hits or misses. In our example,

Longest run of shots hit = 4 (this occurred more than once)
Longest run of shots missed = 5

SECTION 4.2

OVERVIEW

The description of a random phenomenon begins with the **sample space,** which is the list of all possible outcomes. A set of outcomes is called an **event.** Once we have determined the sample space, a **probability model** tells us how to assign probabilities to the various events that can occur. There are four basic rules that probabilities must satisfy.

- Any probability is a number between 0 and 1.
- All possible outcomes together must have probability 1.
- The probability that an event does not occur is 1 minus the probability that the event occurs.
- If two events have no outcomes in common, the probability that one or the other occurs is the sum of their individual probabilities.

In a sample space with a finite number of outcomes, probabilities are assigned to the individual outcomes and the probability of any event is the sum of the probabilities of the outcomes that it contains. In some special cases, the outcomes are all **equally likely** and the probability of any event A is just computed as

$$P(A) = (\text{number of outcomes in } A)/(\text{number of outcomes in } S)$$

Events are **disjoint** if they have no outcomes in common. In this special case the probability that one or the other event occurs is the sum of their individual probabilities. This is the addition rule for disjoint events, namely

$$P(A \text{ or } B) = P(A) + P(B)$$

Events are **independent** if knowledge that one event has occurred does not alter the probability that the second event occurs. The mathematical definition of independence leads to the **multiplication rule** for independent events. If A and B are independent, then

$$P(A \text{ and } B) = P(A)P(B)$$

In any particular problem we can use this definition to check if two events are independent by seeing if the probabilities multiply according to the definition. However, most of the time, independence is assumed as part of the probability model. The four basic rules, plus the multiplication rule, allow us to compute the probabilities of events in many random phenomena.

Many students confuse independent and disjoint events once they have seen both definitions. Remember, disjoint events have no outcomes in common, and when two events are disjoint, you can compute $P(A \text{ or } B) = P(A) + P(B)$ in this special case. The probability being computed is that one or the other event occurs. Disjoint events cannot be independent since once we know that A has occurred, then the probability of B occurring becomes 0 (B cannot have occurred as well—this is the meaning of "disjoint"). The multiplication rule can be used to compute the probability that two events occur *simultaneously*, $P(A \text{ and } B) = P(A)P(B)$, in the special case of independence.

GUIDED SOLUTIONS

Exercise 4.19

KEY CONCEPTS: Sample space

One of the main difficulties encountered when describing the sample space is finding some notation to express your ideas formally. Following the text, our general format is $S = \{ \quad \}$, where a description of the outcomes in the sample space is included within the braces.

In this example, we can let L_1 correspond to clicking on link 1, L_2 correspond to clicking on link 2, and so forth. In addition let E correspond to exiting the Web page. With this notation, describe the sample space for the outcome of a visitor to your Web page.

$S =$ ________________________

Exercise 4.21

KEY CONCEPTS: Applying the probability rules

(a) Since these are the only blood types, what has to be true about the sum of the probabilities for the different types? Use this to find $P(\text{O})$.

(b) What's true about the events O and B blood type? Which probability rule do we follow? (Don't be confused by the wording in the problem, which says "people with blood types O <u>and</u> B." In the context of this problem and the language of probability we are using, it really means "O or B." There are no people with blood types O and B.)

Exercise 4.30

KEY CONCEPTS: Independent events

Although independence is often assumed in setting up a probability model, in other cases we must use the formal definition to determine if two events are independent. In order to determine if the events A = Hispanic and B = white are independent, we must see if they satisfy the multiplication rule. This requires that we evaluate $P(\text{A})$, $P(B)$, and $P(A \text{ and } B)$. To evaluate $P(A)$ we need to add up the proportions of the population corresponding to each race that are also Hispanic.

$$P(A) = 0.000 + 0.003 + 0.060 + 0.062 = 0.125$$

Now evaluate $P(B)$ and $P(A \text{ and } B)$ on your own, and determine if the multiplication rule is satisfied for these events.

$P(B) =$ ____________

$P(A \text{ and } B) =$ ____________

$P(A)P(B) =$ ____________

Exercise 4.31

KEY CONCEPTS: Equally likely outcomes, probabilities of events

(a) The sample space is $S = \{00, 0, 1, 2, 3, \cdots, 36\}$, corresponding to the numbers in the different slots. If a random phenomenon has k possible outcomes, all equally likely, then each individual outcome has probability $1/k$. Use this to determine the probability of landing in any one slot.

(b) If the random phenomenon has equally likely outcomes, then the probability of any event A is

(number of outcomes in A) / (number of outcomes in S)

If A is the event "winning" (or landing on a red slot), how many outcomes are in A? Use this to determine the probability of winning when betting on red.

(c) How many outcomes in S are included in the column bet? Use this to determine the probability of winning with a column bet.

Exercise 4.39

KEY CONCEPTS: Multiplication rule for independent events

(a) The three years are independent. If U indicates a year for the price being up and D indicates a year for the price being down, you need to compute $P(UUU)$.

(b) Since the events are independent, what happens in the first two years does not affect the probability of going up or down in the third year. What's the probability of the price going down in any given year?

(c) This problem must be set up carefully and done in steps.

Step 1. Write the event of interest in terms of simpler outcomes. How would you write $P(UU \text{ or } DD)$ in terms of $P(UU)$ and $P(DD)$?

P(moves in the same direction in the next two years) = $P(UU \text{ or } DD)$ = _______________

Step 2. Evaluate $P(UU)$ and $P(DD)$ and substitute your answer in the expression from Step 1.

Exercise 4.43

KEY CONCEPTS: Independence, multiplication rule

a) The possible alleles inherited are B and B, B and O, and O and O. What blood types do these inherited alleles result in?

b) Let N_O and N_B correspond to the events that allele O or B is inherited from Nancy, respectively, and D_O and D_B correspond to the events that allele O or B is inherited from David. N_O and N_B each has probability 0.5 and similarly for D_O and D_B.

P(child has type O) = $P(N_O \text{ and } D_O)$ =

What rule allows you to multiply the probabilities?

What is the probability that the child has type B blood?

COMPLETE SOLUTIONS

Exercise 4.19

Using the notation E corresponding to exiting the Web page, L_1 correspond to clicking on link 1, L_2 correspond to clicking on link 2, and so forth, the sample space is given by S = {L_1, L_2, L_3, L_4, L_5, E}. This sample space describes all possible outcomes for a visitor to your Web page.

Exercise 4.21

(a) The probabilities for the different blood types must add to 1. The sum of the probabilities for blood types A, B, and AB is 0.40 + 0.11 + 0.04 = 0.55. Subtracting this from 1 tells us that the probability of the remaining type, O, must be 0.45.

(b) Maria can receive transfusions from people with blood types O or B. Since a person cannot have both of these blood types, they are disjoint. The calculation follows the text's probability rule 3, which says $P(\text{O or B}) = P(\text{O}) + P(\text{B}) = 0.45 + 0.11 = 0.56$.

Exercise 4.30

In the Guided Solutions it is shown that $P(A) = 0.125$. $P(B)$ is computed as

$$P(B) = 0.060 + 0.691 = 0.751$$

The probability of being both white and Hispanic corresponds to the single entry in the column labeled "Hispanic" and the row labeled "white." Reading the entry in the table gives $P(A \text{ and } B) = 0.060$. Since $P(A)P(B) = 0.125 \times 0.751 = 0.094$ is not equal to $P(A \text{ and } B)$, the multiplication rule is not satisfied and the events are not independent.

Exercise 4.31

(a) S has $k = 38$ equally likely outcomes. Thus, the probability of landing on any slot is $1/38 = 0.0263$.

(b) The event A, landing on a red slot, corresponds to 18 of the equally likely outcomes, so that the probability of winning when betting on red is $18/38 = 0.4737$.

(c) There are 12 multiples of 3, so the event A consists of 12 equally likely outcomes. The probability of winning with this column bet is $12/38 = 0.3158$.

Exercise 4.39

(a) $P(UUU) = (0.65)^3 = 0.2746$

(b) The probability of the price being down in any given year is $1 - 0.65 = 0.35$. Since the years are independent, the probability of the price being down in the third year is 0.35, regardless of what has happened in the first two years.

(c) P(moves in the same direction in the next two years) = $P(UU$ or $DD) = P(UU) + P(DD)$, since the events UU and DD are disjoint. Using the independence of two successive years, $P(UU) = (0.65)^2 = 0.4225$, and $P(DD) = (0.35)^2 = 0.1225$. Putting this together,

$$P(\text{moves in the same direction in the next two years}) = 0.4225 + 0.1225 = 0.5450$$

Exercise 4.43

a) The possible alleles inherited are B and B, B and O, and O and O. The alleles B and B and B and O both result in a blood type of B for a child. The alleles O and O result in a blood type of O for a child. So the two blood types their children can have are B and O.

b) Let N_O and N_B correspond to the events that allele O or B is inherited from Nancy, respectively, and D_O and D_B correspond to the events that allele O or B is inherited from David. N_O and N_B each have probability 0.5 and similarly for D_O and D_B.

$$P(\text{child has type O}) = P(N_O \text{ and } D_O) = 0.5 \times 0.5 = 0.25$$

You multiply the probabilities because we inherit alleles independently from our mother and father. Since the child must have blood type B or O, the P(child has type B) = 1 – P(child has type O) = 1 – 0.25 = 0.75.

SECTION 4.3

OVERVIEW

A **random variable** is a variable whose value is a numerical outcome of a random phenomenon. The restriction to numerical outcomes makes the description of the probability model simpler and allows us to begin to look at some further properties of probability models in a unified way. If we toss a coin three times and record the sequence of heads and tails, then an example of an outcome would be HTH, which would not correspond directly to a random variable. On the other hand, if we were only keeping track of the number of heads on the three tosses, then the outcome of the experiment would be 0, 1, 2, or 3 and would correspond to the value of the random variable X = number of heads.

The two types of random variables we will encounter are **discrete** and **continuous** random variables. The **probability distribution** of a random variable tells us about the possible values of X and how to assign probabilities to these values. A discrete random variable has a finite number of values, and the probability distribution is a list of the possible values of X and the probabilities assigned to these values. The probability distribution can be given in a table or using a **probability histogram**. For any event described in terms of X, the probability of the event is just the sum of the probabilities of the values of X included in the event.

A continuous random variable takes all values in some interval of numbers. Probabilities of events are determined using a **density curve**. The probability of any event is the area under the curve corresponding to the values that make up the event. For density curves that involve regular shapes such as rectangles or triangles, we can compute probabilities of events using simple geometrical arguments. The **normal distribution** is another example of a continuous probability distribution, and probabilities of events for normal random variables are computed by standardizing and referring to Table A as was done in Section 1.3.

GUIDED SOLUTIONS

Exercise 4.55

KEY CONCEPTS: Discrete random variables, computing probabilities

(a) Write the event in terms of a probability about the random variable X. Although you can figure out the answer without doing this, it's good practice to start using the notation for random variables. To find the probability that "the unit has 6 or more rooms," add the appropriate probabilities given in Exercise 4.53 which provides the distribution of X. (Be sure to use the row for owned units).

(b) Express the event in words and compute its probability using the distribution of X in Exercise 4.53. How is this event different from the event in part (a)?

(c) You should have different answers in part (a) and part (b). What fact about discrete random variables does this illustrate? If X had a continuous distribution, would the events in (a) and (b) have the same probability?

Exercise 4.60

KEY CONCEPTS: Finding the probability distribution of a random variable

(a) The probability of a randomly selected student opposing the funding of interest groups is 0.4 and the probability of favoring it is 0.6. The opinions of different students sampled are independent of each other. You can use the multiplication rule to find

P(A supports, B supports, and C opposes) =

(b) It is easiest to do this by making a table to keep track of the calculations. The first entry is given below. There should be eight lines to the table when you're done. If you've done the calculations correctly, what should be true about the eight probabilities? Don't worry about the column labeled "value of *X*" for now. It will not be needed until part (c).

A	B	C	Probability	Value of X
support	support	support	$(0.6)^3 = 0.216$	

(c) For each committee listed in the table in (b), find the associated value of *X*. For the first row, 0 people oppose the funding of interest groups, so the value of *X* is 0 for a committee with these views. The values of *X* that can occur are 0, 1, 2, and 3. To find the probability that *X* takes any of these values, just add up the probabilities of the committees with that value of *X*. Fill in the table below with your values and make sure that the probabilities sum to 1.

Value of X	Probability

(d) If a majority opposes funding, how many people on the committee would have to oppose funding? What does this say about *X*? Now use the table you constructed in (c) to evaluate this probability.

Exercise 4.63

KEY CONCEPTS: Continuous random variables, computing probabilities

(a) The graph of the density curve is given below. The property you are given is that the density has a constant height between 0 and 2. Since the area under the density curve must be equal to 1, what is the height? Recall that the area of a rectangle is length × height.

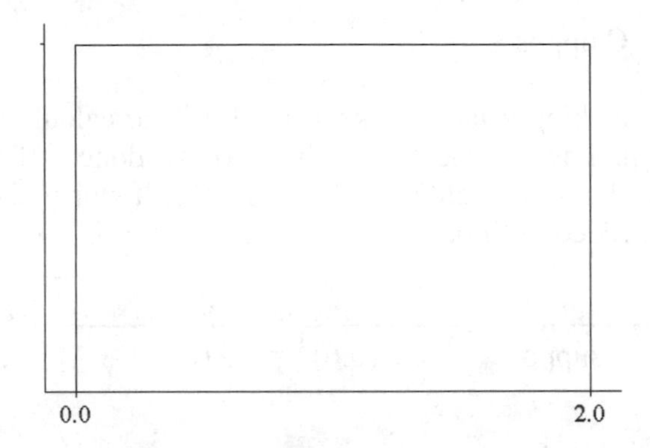

(b) As with finding areas under normal curves, it helps to draw a sketch of the density curve that includes the area that corresponds to the probability that you need to evaluate. In this part you need to find $P(Y \leq 1.5)$, when Y is a random number between 0 and 2. The density curve with the area corresponding to this probability is given below. Since it is a rectangular region, the area corresponds to the length × height = $1.5 \times 0.5 = 0.75$, which is the $P(Y \leq 1.5)$. Remember, for continuous densities, $P(Y \leq 1.5) = P(Y < 1.5)$.

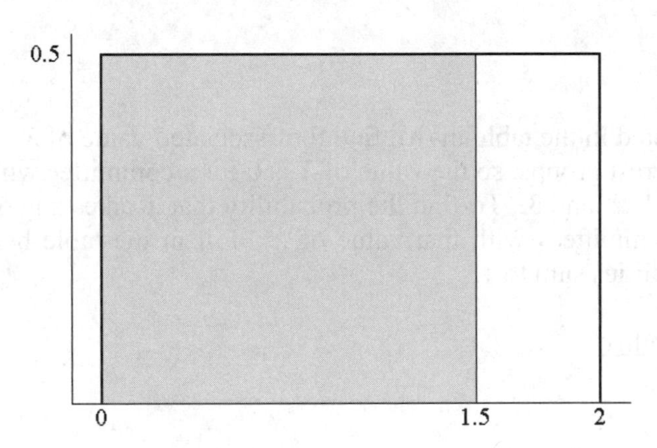

(c) Sketch the density curve and the area you need below.

(d) Sketch the density curve and the area you need below.

Exercise 4.65

KEY CONCEPTS: Probabilities for a sample mean

Finding probabilities associated with a sample mean $\bar{x}$ when we know the mean and standard deviation of the sampling distribution requires first standardizing to a *z*-score so that we can refer to the table for the standard normal distribution. In this example, $\bar{x}$ has a mean of 9.0 and the standard deviation is given as 0.075. If you are still uncomfortable doing this type of problem, it is best to continue to draw a picture of a normal curve and the required area as we did in Section 1.3. Otherwise, you can just follow the method of Example 4.26 in this chapter. Below is a picture of a normal curve with a mean of 9.0; the standard deviation is given as 0.075, and the required area is shaded. Now, proceed as in the problems in Section 1.3 to find the area of the shaded portion.

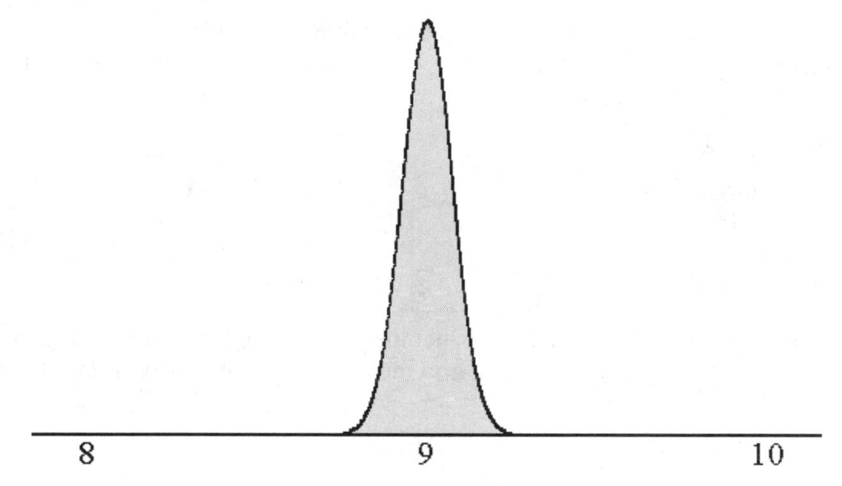

$P(8 < \bar{x} < 10) =$

COMPLETE SOLUTIONS

Exercise 4.55

(a) $P(X \geq 6) = 0.224 + 0.197 + 0.149 + 0.053 + 0.035 = 0.658$

(b) The event $\{X > 6\}$ indicates that "the unit has *more than* 6 rooms." The probability is computed as

$P(X > 6) = 0.197 + 0.149 + 0.053 + 0.035 = 0.434$

(c) The answers are different because for discrete distributions the probability of an individual outcome is not necessarily zero. In this case, the probability of the individual outcome 6 is not zero, so that $P(X > 6)$ does not have the same value as $P(X \geq 6)$. (**Note**: For this distribution of X, $P(X > 4.5) = P(X \geq 4.5)$ since the outcome 4.5 has zero probability.)

Exercise 4.60

(a) *P*(A supports, B supports, and C opposes)
= *P*(A supports) *P*(B supports) *P*(C opposes) = (0.6)(0.6)(0.4) = 0.144

(b)

A	B	C	Probability		Value of X
support	support	support	$(0.6)^3$	$= 0.216$	0
support	support	oppose	$(0.6)^2(0.4)$	$= 0.144$	1
support	oppose	support	$(0.6)^2(0.4)$	$= 0.144$	1
oppose	support	support	$(0.6)^2(0.4)$	$= 0.144$	1
support	oppose	oppose	$(0.6)(0.4)^2$	$= 0.096$	2
oppose	support	oppose	$(0.6)(0.4)^2$	$= 0.096$	2
oppose	oppose	support	$(0.6)(0.4)^2$	$= 0.096$	2
oppose	oppose	oppose	$(0.4)^3$	$= 0.064$	3
			Total	$= 1.000$	

(c) The values of X are given in the table in (b). The possible values of X are 0, 1, 2, and 3. To find the probability that X takes any value, just add up the probabilities of the committees with that value of X. For example, $P(X = 2) = 3(0.096) = 0.288$.

Value of X	Probability
0	0.216
1	0.432
2	0.288
3	0.064

(d) "A majority oppose" means that either 2 or 3 members of the board oppose. If 2 oppose then $X = 2$, and if 3 oppose then $X = 3$. So the event is $X \geq 2$ and the required probability is $P(X \geq 2) = 0.288 + 0.064 = 0.352$.

Exercise 4.63

(a) The area is the length $\times$ height. Since the area is 1 and we know the length is 2, you must solve the equation $2 \times$ height $= 1$. Thus, the height must be equal to 1/2 or 0.5.

(b) See the Guided Solutions.

(c) The shaded area is $P(0.6 < Y < 1.7) = (1.1)(0.5) = 0.55$.

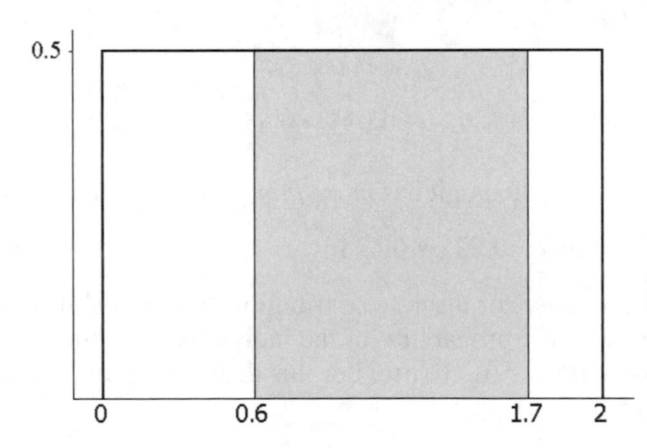

d) The shaded area is $P(Y \geq 0.9) = (1.1)(0.5) = 0.55$.

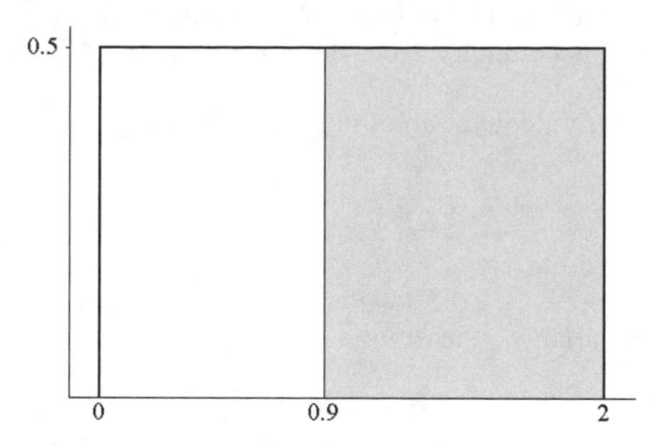

Exercise 4.65

$$P(8 < \overline{x} < 10) = P\left(\frac{8-9}{0.075} < \frac{\overline{x}-9}{0.075} < \frac{10-9}{0.075}\right)$$

$$= P(-13.33 < Z < 13.33) \doteq 1$$

The chance that $\overline{x}$ will estimate μ to within ± 1 is essentially 1.

SECTION 4.4

OVERVIEW

In Chapter 1 we introduced the concept of the distribution of a set of numbers or data. The distribution describes the different values in the set and the frequency or relative frequency with which those values occur. The mean of the numbers is a measure of the center of the distribution and the standard deviation is a measure of the variability or spread. These concepts are also used to describe features of a random variable X. The probability distribution of a random variable indicates the possible values of the random variable and the probability (relative frequency in repeated observations) with which they occur. The **mean** μ_X of a random variable X describes the center or balance point of the probability distribution or density curve of X. If X is a discrete random variable having possible values $x_1, x_2, \ldots, x_k$ with corresponding probabilities $p_1, p_2, \ldots, p_k$, the mean μ_X is the average of the possible values weighted by the corresponding probabilities, i.e.,

$$\mu_X = x_1p_1 + x_2p_2 + \ldots + x_kp_k$$

The mean of a continuous random variable is computed from the density curve but computations require more advanced mathematics. The law of large numbers relates the mean of a set of data to the mean of a random variable and says that the average of the values of X observed in many trials approaches μ_X.

The **variance** σ_X^2 of a random variable X is the average squared deviation of the values of X from their mean. For a discrete random variable

$$\sigma_X^2 = (x_1 - \mu_X)^2p_1 + (x_2 - \mu_X)^2p_2 + \ldots + (x_k - \mu_X)^2p_k$$

The **standard deviation σ_X** is the positive square root of the variance. The standard deviation measures the variability of the distribution of the random variable *X* about its mean. The variance of a continuous random variable, like the mean, is computed from the density curve. Again, computations require more advanced mathematics.

The mean and variances of random variables obey the following rules. If *a* and *b* are fixed numbers then

$$\mu_{a+bX} = a + b\mu_X$$

$$\sigma^2_{a+bX} = b^2\sigma^2_X$$

If *X* and *Y* are any two random variables, then

$$\mu_{X+Y} = \mu_X + \mu_Y$$

If *X* and *Y* are independent random variables, then

$$\sigma^2_{X+Y} = \sigma^2_X + \sigma^2_Y$$

$$\sigma^2_{X-Y} = \sigma^2_X + \sigma^2_Y$$

If *X* and *Y* have correlation ρ, then the general addition rule for variances of random variables is

$$\sigma^2_{X+Y} = \sigma^2_X + \sigma^2_Y + 2\rho\sigma_X\sigma_Y$$

$$\sigma^2_{X-Y} = \sigma^2_X + \sigma^2_Y - 2\rho\sigma_X\sigma_Y$$

GUIDED SOLUTIONS

Exercise 4.73

KEY CONCEPTS: Distributions, mean

Recall that the average (mean) of *X* is computed using the formula

$$\mu_X = x_1p_1 + x_2p_2 + \ldots + x_kp_k$$

where the values of x_i are the values for the different grades and the p_i are the probabilities of these values. Do the calculations to evaluate the mean in the space below.

$\mu_X =$

Exercise 4.79

KEY CONCEPTS: Means and variances of sums, calculating mean and standard deviation of a distribution

(a) The distribution of *X* is

Number of heads (x)	0	1
Probability (p)	0.5	0.5

Recall that the average (mean) of X = number of heads is computed using the formula

$$\mu_X = x_1p_1 + x_2p_2 + \ldots + x_kp_k$$

where the values of x_i and the p_i are given in the preceding table.

μ_X = ____________________

Once you have calculated the mean μ_X, the general formula for the variance is

$$\sigma_X^2 = (x_1 - \mu_X)^2p_1 + (x_2 - \mu_X)^2p_2 + \ldots + (x_k - \mu_X)^2p_k$$

After you have calculated the variance, remember to take the square root to obtain the standard deviation. If you are having trouble with these formulas, review Example 4.33 in the text.

σ_X^2 = ____________________

σ_X = ____________________

(b) Let X_1 be the number of heads on the first toss, X_2 be the number of heads on the second toss, X_3 be the number of heads on the third toss, and X_4 be the number of heads on the fourth toss. Then

$$Y = X_1 + X_2 + X_3 + X_4$$

is the number of heads on the four tosses. Use the rules for means and variances of sums to find the mean and variance of Y. Remember that X_1, X_2, X_3, and X_4 are independent.

$\mu_Y = \mu_{X_1+X_2+X_3+X_4} = \mu_{X_1} + \mu_{X_2} + \mu_{X_3} + \mu_{X_4} =$

$\sigma_Y^2 = \sigma_{X_1+X_2+X_3+X_4}^2 = \sigma_{X_1}^2 + \sigma_{X_2}^2 + \sigma_{X_3}^2 + \sigma_{X_4}^2 =$

$\sigma_Y =$

(c) From Example 4.23, the distribution of Y is

Number of heads (y)	0	1	2	3	4
Probability (p)	0.0625	0.25	0.375	0.25	0.0625

Find the mean and standard deviation of this distribution directly and compare with your answer in (b). Which method is simpler?

$\mu_Y =$

$\sigma_Y^2 =$

$\sigma_Y =$

Exercise 4.81

KEY CONCEPTS: Rules for means and variances

Recall that if X, Y, and Z are any three random variables, then

$$\mu_{X+Y+Z} = \mu_X + \mu_Y + \mu_Z$$

In this problem, let X be the length of the bearing on the left, Y be length of the rod, and Z be the length of the bearing on the right. Then the total length of the assembly is $X + Y + Z$. Use the information in the problem and the rule for the mean of $X + Y + Z$ to evaluate μ_{X+Y+Z}, the mean total length of the assembly.

$\mu_{X+Y+Z} =$

In order to find the standard deviation of the total length you must first find the variance. When X, Y, and Z are independent random variables, then the formula for the variance of $X + Y + Z$ follows Rule 2 on page 282 for variances and is

$$\sigma^2_{X+Y+Z} = \sigma^2_X + \sigma^2_Y + \sigma^2_Z$$

Using the information given in the problem, evaluate both the variance and the standard deviation. The previous formula requires the variances of the three components, but the problem gives the standard deviations of these components so be sure to first convert the standard deviations to variances before using the formula for σ^2_{X+Y+Z}.

$\sigma^2_{X+Y+Z} =$

$\sigma_{X+Y+Z} =$

Exercise 4.89

KEY CONCEPTS: Mean of a random variable

(a) You are given the probabilities that he will die in each year over the next five years. The probability of not dying is the probability of age at death being 30 years or greater. Compute the probability of not dying in the next five years by filling in the blank in the table below.

Age at death (x)	25	26	27	28	29	≥ 30
Probability (p)	0.00039	0.00044	0.00051	0.00057	0.00060	

(b) Recall that the average (mean) of X = cash intake is computed using the formula

$$\mu_X = x_1p_1 + x_2p_2 + \ldots + x_kp_k$$

The values of x_i are the cash intakes and the p_i are given in the table in part (a). You need to remember that if the man does not die in the next five years, then the cash intake is positive, while the losses correspond to negative cash intakes. Also, the probabilities in the distribution must sum to 1, so you cannot leave out the category ≥ 30.

Age at death	25	26	27	28	29	≥ 30
Cash intake (x)	–$99,825	–$99,650	–$99,475	–$99,300	–$99,125	$875

Now use the formula to compute μ_X. The first five values of X are negative (since there is a loss) and the last is positive, so be careful with the signs of the products when computing the mean of X.

$\mu_X =$

Exercise 4.91

KEY CONCEPTS: Rules for means and variances

This exercise is very similar to Example 4.38 in the text; if you are having difficulty with this exercise, you should review Example 4.38. The portfolio consists of 80% 500 Index Fund (W) and 20% Diversified International Fund (Y). Letting the rate of return on this portfolio be R, first write R in terms of W and Y.

$R =$

Using the information in the problem, fill in the properties of the returns on the 500 Index Fund and the Diversified International Fund in the spaces provided.

W = annual return on the 500 Index Fund $\mu_W =$ $\sigma_W =$

Y = annual return on the Diversified International Fund $\mu_Y =$ $\sigma_Y =$

Correlation between W and Y $\rho_{WY} =$

The return on the portfolio of 80% 500 Index Fund (W) and 20% Diversified International Fund (Y) has a mean of

$\mu_R =$

To find the variance of the portfolio you need to combine Rules 1 and 3 for variances as in Example 4.38 of the text.

$\sigma_R^2 =$

Now you can find the standard deviation of the return. This portfolio has a similar mean return to either the 500 Index Fund or the Diversified International Fund. In what sense is it less risky?

$\sigma_R =$

COMPLETE SOLUTIONS

Exercise 4.73

The average grade in this course is

$$\mu_X = 0.05(0) + 0.04(1) + 0.20(2) + 0.40(3) + 0.31(4) = 2.88.$$

Exercise 4.79

(a) We compute

$$\mu_X = 0(0.5) + 1(0.5) = 0.5$$

$$\sigma_X^2 = (0 - 0.5)^2(0.5) + (1 - 0.5)^2(0.5)$$

$$= 0.125 + 0.125$$

$$= 0.25$$

$$\sigma_X = \sqrt{0.25} = 0.5$$

(b) From part (a) we know $\mu_{X_1} = \mu_{X_2} = \mu_{X_3} = \mu_{X_4} = 0.5$ and $\sigma_{X_1}^2 = \sigma_{X_2}^2 = \sigma_{X_3}^2 = \sigma_{X_4}^2 = 0.25$. Substituting these values in the equations for the mean and variance of Y gives

$$\mu_Y = \mu_{X_1+X_2+X_3+X_4} = \mu_{X_1} + \mu_{X_2} + \mu_{X_3} + \mu_{X_4} = 0.5 + 0.5 + 0.5 + 0.5 = 2.0$$
$$\sigma_Y^2 = \sigma_{X_1+X_2+X_3+X_4}^2 = \sigma_{X_1}^2 + \sigma_{X_2}^2 + \sigma_{X_3}^2 + \sigma_{X_4}^2 = 0.25 + 0.25 + 0.25 + 0.25 = 1.0$$

$$\sigma_Y = 1.0$$

c) We compute

$$\mu_Y = 0(0.0625) + 1(0.25) + 2(0.375) + 3(0.25) + 4(0.0625)$$

$$= 0 + 0.25 + 0.75 + 0.75 + 0.25$$

$$= 2.0$$

$$\sigma_Y^2 = (0 - 2)^2(0.0625) + (1 - 2)^2(0.25) + (2 - 2)^2(0.375) + (3 - 2)^2(0.25) + (4 - 2)^2(0.0625)$$

$$= 0.25 + 0.25 + 0 + 0.25 + 0.25$$

$$= 1.0$$

$$\sigma_Y = \sqrt{1.0} = 1.0$$

The same answers are obtained in parts (b) and (c) but the technique used in part (b) is much simpler, particularly if we increase the number of tosses.

Exercise 4.81

KEY CONCEPTS: Rules for means and variances

We are told that $\mu_X = 2$ cm, $\mu_Y = 12$ cm, and $\mu_Z = 2$ cm, where *X* and *Z* are the lengths of the two bearings and *Y* is the length of the rod. Substituting these values in the formula for the mean of the total length of the assembly gives

$$\mu_{X+Y+Z} = 2 + 12 + 2 = 16\text{cm}.$$

Using the information given in the problem we have $\sigma_X^2 = (0.001)^2$, $\sigma_Y^2 = (0.004)^2$, and $\sigma_Z^2 = (0.001)^2$. Substituting these values in the formula for the variance of the total length of the assembly gives

$$\sigma_{X+Y+Z}^2 = (0.001)^2 + (0.004)^2 + (0.001)^2 = 0.000018$$

and the standard deviation as

$$\sigma_{X+Y+Z} = \sqrt{0.000018} = 0.00424$$

Exercise 4.89

(a) The probabilities of dying in the next five years add to

$$0.00039 + 0.00044 + 0.00051 + 0.00057 + 0.00060 = 0.00251$$

so the probability that a 25-year-old man doe not die in the next 5 years is $1 - 0.00251 = 0.99749$.

(b) We calculate

$$\begin{aligned}\mu_X &= 0.00039(-99825) + 0.00044(-99650) + 0.00051(-99475) \\ &\quad + 0.00057(-99300) + 0.00060(-99125) + 0.99749(875) \\ &= -38.932 - 43.846 - 50.732 - 56.601 - 59.475 + 872.804 \\ &= 623.218\end{aligned}$$

Not surprisingly, the mean is positive so the insurance company expects to make a little over \$623 per policy that it sells. Like any form of gambling, in the long run the insurance companies will make money despite an occasional large payout.

Exercise 4.91

Letting the rate of return on this portfolio be *R*, this can be written in terms of *W* and *Y*.

$$R = 0.8W + 0.2Y$$

The returns on the 500 Index Fund and the Diversified International Fund have the following properties:

W = annual return on the 500 Index Fund $\mu_W = 11.12$ $\sigma_W = 17.46$

Y = annual return on the Diversified International Fund $\mu_Y = 11.10$ $\sigma_Y = 15.62$

Correlation between W and Y $\rho_{WY} = 0.56$

The return on the portfolio of 80% 500 Index Fund (W) and 20% Diversified International Fund (Y) has a mean of

$$\mu_R = 0.8\,\mu_W + 0.2\,\mu_y = (0.8 \times 11.12) + (0.2 \times 11.10) = 11.116\%$$

To find the variance of the portfolio you need to combine Rules 1 and 3 for variances as in Example 4.38 of the text.

$$\begin{aligned}\sigma_R^2 &= \sigma_{0.8W}^2 + \sigma_{0.2Y}^2 + 2\rho_{WY}(0.8\times\sigma_W)(0.2\times\sigma_Y)\\ &= (0.8)^2(17.46)^2 + (0.2)^2(15.62)^2 + (2)(0.56)(0.8\times17.46)(0.2\times15.62) = 253.74\end{aligned}$$

$\sigma_R = 15.93\%$

The portfolio has almost the rate of return of the 500 Index Fund with a rate of risk almost as low as the Diversified International Fund.

SECTION 4.5

OVERVIEW

This section discusses a number of basic concepts and rules that are used to calculate probabilities of complex events. The **complement** A^c of an event A contains all the outcomes in the sample space that are not in A. It is the "opposite" of A. The **union** of two events A and B contains all outcomes in A, in B, or in both. The union is sometimes referred to as the event A or B. The **intersection** of two events A and B contains all outcomes that are in both A and B simultaneously. The intersection is sometimes referred to as the event A and B. We say two events A and B are **disjoint** if they have no outcomes in common.

The **conditional probability** of an event B given an event A is denoted by $P(B|A)$ and is defined by

$$P(B \mid A) = \frac{P(A \text{ and } B)}{P(A)}$$

when $P(A) > 0$. In practice it can often be determined directly from the information given in a problem. Two events A and B are **independent** if $P(B|A) = P(B)$.

Other general rules of elementary probability are

- Legitimate values: $0 \le P(A) \le 1$ for any event A
- Total probability: $P(S) = 1$, where S denotes the sample space.
- Complement rule: $P(A^c) = 1 - P(A)$
- Addition rule: $P(A \text{ or } B) = P(A) + P(B) - P(A \text{ and } B)$
- Multiplication rule: $P(A \text{ and } B) = P(A)P(B|A)$
- Bayes's rule: $P(A \mid B) = \dfrac{P(B \mid A)P(A)}{P(B \mid A)P(A) + P(B \mid A^c)P(A^c)}$ provided $0 < P(A), P(B) < 1$
- For disjoint events: $P(A \text{ and } B) = 0$ and so $P(A \text{ or } B) = P(A) + P(B)$
- For independent events: $P(A \text{ and } B) = P(A)P(B)$

In problems with several stages, it is helpful to draw a tree diagram to guide you in the use of the multiplication and addition rules.

GUIDED SOLUTIONS

Exercise 4.109

KEY CONCEPTS: Events, probability rules, Venn diagrams

The notation defined for the events in this problem are A = car and B = imported. The events described in parts (a) and (b) of this problem can be defined in terms of A and B. You are told that $P(A^c) = 0.69$, $P(B^c) = 0.78$, and $P(A^c \text{ and } B^c) = 0.55$.

(a) The event "the vehicle is a light truck" corresponds to what event in terms of A and B? If necessary, use the probability rules to evaluate the probability that the vehicle is a light truck.

(b) "The vehicle is an imported car" corresponds to the event "A and B." From the addition rule

$$P(A \text{ or } B) = P(A) + P(B) - P(A \text{ and } B), \text{ or equivalently } P(A \text{ and } B) = P(A) + P(B) - P(A \text{ or } B).$$

You can find both $P(A)$ and $P(B)$ from the information given using the complement rule.

$P(A) = 1 - \text{P}(A^c) =$

$P(B) = 1 - \text{P}(B^c) =$

Now first verify the following relationship is true using a Venn diagram

$P(A \text{ or } B) = 1 - P(A^c \text{ and } B^c) =$

and using the answers you have obtained compute

$P(A \text{ and } B) = P(A) + P(B) - P(A \text{ or } B) =$

Using a Venn diagram may lead you to another approach to get to the answer which is simpler for you.

Exercise 4.111

KEY CONCEPTS: Conditional probability, independence

(a) You are asked to evaluate *P*(vehicle is a light truck | vehicle is imported) = $P(A^c \mid B)$. Using the definition of conditional probability you need to evaluate

$$P(A^c \mid B) = \frac{P(A^c \text{ and } B)}{P(B)} =$$

For the numerator, first verify using a Venn diagram that

$P(B) = P(A \text{ and } B) + P(A^c \text{ and } B)$, or equivalently that $P(A^c \text{ and } B) = P(B) - P(A \text{ and } B)$

$P(A^c \text{ and } B) = P(B) - P(A \text{ and } B) =$

You should now be able to complete the problem.

(b) If the events "vehicle is a light truck" and "vehicle is imported" are independent, what should be true about the two probabilities *P*(vehicle is a light truck) and *P*(vehicle is a light truck | vehicle is imported)? Are these two events independent?

Exercise 4.117

KEY CONCEPTS: Conditional probabilities and the multiplication rule

The table in Exercise 4.116 is reproduced below to assist you.

	Bachelor's	Master's	Professional	Doctorate	Total
Female	933	402	51	26	1412
Male	661	260	44	26	991
Total	1594	662	95	52	2403

(a) You can calculate this probability directly from the table. All degree recipients in the table are equally likely to be selected (that is what it means to select a degree recipient at random), so that the fraction of the degree recipients in the table that are men is the desired probability. How many degree recipients are men? Where do you find this in the table? What is the total number of degree recipients represented in the table? Use these numbers to compute the desired fraction.

(b) This probability can also be calculated directly from the table. Since this is a conditional probability (i.e., this is a probability given that the degree recipient is a man), we restrict ourselves only to degree recipients that are men. The desired probability is then the fraction of these men that received a bachelor's degree. Use the appropriate entries in the table to compute this fraction.

(c) Recall that the multiplication rule says

P(degree recipient is both "male" and "received a bachelor's")

= P(degree recipient is male)P(degree recipient received bachelor's | degree recipient is male)

Using the answers in (a) and (b), calculate this probability.

The number of degree recipients that are both "male" and "received a bachelor's" can be read directly from the table. What is this number? What fraction of the total number of degree recipients represented in the table is this number? This should agree with the probability you calculated using the multiplication rule.

Exercise 4.127

KEY CONCEPTS: Multiplication rules and conditional probability

Let the event C correspond to a person of European ancestry carrying an abnormal CF gene and the event + correspond to the CF20m test being positive. We are given the following probabilities.

$$P(C) = 0.04$$

$$P(+ \mid C) = 0.90$$

$$P(+ \mid C^c) = 0.00$$

You are asked to compute the conditional probability of the event "Jason is a carrier" given that "Jason tests positive." Before trying to apply any formulas think about what the problem is asking. Can a person who is not a carrier test positive? What must be the probability that Jason is a carrier given that he tests positive? In the Guided Solutions we provide the formulas to formally verify the answer.

Exercise 4.128

KEY CONCEPTS: Bayes's rule

Before beginning, note that the complement rule also applies to conditional probabilities, namely for any events A and B,

$$P(A^c \mid B) = 1 - P(A \mid B)$$

You will need to use the complement rule for conditional probabilities in this and other exercises.
Let H be the event "Husband is a carrier," W be the event "Wife is a carrier," and C be the event "First Child has CF". Fill in the following probabilities which are given in the problem, or use the complement rule for conditional probabilities.

$P(H) =$

$P(W) =$

$P(C \mid W) =$ $\qquad$ $P(C^c \mid W) =$

$P(C \mid W^c) =$ $\qquad$ $P(C^c \mid W^c) =$

You are asked to compute the conditional probability of the event "Wife is a carrier" given that "First Child doesn't have CF." or using our notation

$$P(\text{"Wife is a carrier"} \mid \text{"First Child doesn't have CF"}) = P(W \mid C^c) = \frac{P(W \text{ and } C^c)}{P(C^c)}$$

This is an application of Bayes's rule because we are calculating the "reverse" conditional probability from the ones we are given. Although it is an example of Bayes's rule, it is better to work your way through the problem as in Examples 4.47 and 4.48 of the text rather than trying to memorize the rule.

$P(W \text{ and } C^c)$ can be calculated using the multiplication rule and the probabilities you have filled in above.

$P(W \text{ and } C^c) = P(C^c \mid W)\, P(W)$

To find the $P(C^c)$, you need to use the multiplication rule for both probabilities below.

$P(C^c) = P(W \text{ and } C^c) + P(W^c \text{ and } C^c) =$

Use these two calculations to evaluate

$$P(W \mid C^c) = \frac{P(W \text{ and } C^c)}{P(C^c)} =$$

COMPLETE SOLUTIONS

Exercise 4.109

The information given in the problem is $P(A^c) = 0.69$, $P(B^c) = 0.78$ and $P(A^c \text{ and } B^c) = 0.55$.

(a) $P(\text{vehicle is a light truck}) = P(A^c) = 0.69$

(b) "The vehicle is an imported car" corresponds to the event A and B. Following the Guided Solutions we first find

$P(A) = 1 - P(A^c) = 1 - 0.69 = 0.31$

$P(B) = 1 - P(B^c) = 1 - 0.78 = 0.22$

Since we are given $P(A^c \text{ and } B^c) = 0.55$, it follows that

$P(A \text{ or } B) = 1 - P(A^c \text{ and } B^c) = 1 - 0.55 = 0.45$

Now,

$P(A \text{ and } B) = P(A) + P(B) - P(A \text{ or } B) = 0.31 + 0.22 - 0.45 = 0.08$

Exercise 4.111

(a) Using the results from Exercise 4.109,

$P(A^c \text{ and } B) = P(B) - P(A \text{ and } B) = 0.22 - 0.08 = 0.14$, and

$$P(A^c \mid B) = \frac{P(A^c \text{ and } B)}{P(B)} = \frac{0.14}{0.22} = 0.6364$$

(b) The two probabilities *P*(vehicle is a light truck) and *P*(vehicle is a light truck | vehicle is imported) should be equal if the events "vehicle is a car" and "vehicle is imported" are independent. From part (a) of this exercise,

P(vehicle is a light truck | vehicle is imported) = 0.6364

and from part (a) of Exercise 4.109, *P*(vehicle is a light truck) = 0.69. Thus, these two events are not independent.

Exercise 4.117

(a) The number of degree recipients that are men is found at the end of the row labeled "male" and is (in thousands) 991. The total number of degree recipients in the table is in the lower right corner and is (in thousands) 2403. The desired probability is thus

(number of degree recipients that are male)/(number of degree recipients) = 991/2403 = 0.4124

(b) The desired probability is

(number of males who received bachelor's)/(number of males) = 661/991 = 0.6670

(c) Recall that the multiplication rule says

P(degree recipient is both "male" and "received a bachelor's")

= *P*(degree recipient is male)*P*(degree recipient received bachelor's | degree recipient is male)

= (0.4124)(0.6670) = 0.2751

The number of degree recipients that are both "male" and "received a bachelor's" can be read directly from the table and is (in thousands) 661 of the 2403 degree recipients, giving a probability of 661/2403 = 0.2751, which agrees with the answer obtained using the multiplication rule.

Exercise 4.127

We are given the following probabilities.

$$P(C) = 0.04$$

$$P(+ \mid C) = 0.90$$

$$P(+ \mid C^c) = 0.00$$

You are asked to compute the conditional probability of the event "Jason is a carrier" given that "Jason tests positive," or using our notation

$$P(C \mid +) = \frac{P(C \text{ and } +)}{P(+)}$$

Using the multiplication rule,

$$P(C \text{ and } +) = P(C)P(+ \mid C) = 0.04 \times 0.90 = 0.36$$

Now,

$$P(+) = P(+ \text{ and } C) + P(+ \text{ and } C^c) = P(C)P(+ \mid C) + P(C^c)P(+ \mid C^c)$$

$$= (0.04 \times 0.90) + (0.96 \times 0.00) = 0.36,$$

so we see that $P(C \mid +) = 1$, which can be deduced without using the formulas because $P(+ \mid C^c) = 0.00$.

Exercise 4.128

Let H be the event "Husband is a carrier," W be the event "Wife is a carrier," and C be the event "First Child has CF." The following probabilities are given in the problem.

$$P(H) = 1$$

$$P(W) = 2/3$$

$$P(C \mid W) = 1/4 \qquad P(C^c \mid W) = 3/4$$

$$P(C \mid W^c) = 0 \qquad P(C^c \mid W^c) = 1$$

From these probabilities and the multiplication rule we have

$$P(W \text{ and } C^c) = P(C^c \mid W)\, P(W) = 3/4 \times 2/3 = 1/2$$

and

$$P(W^c \text{ and } C^c) = P(C^c \mid W^c)\, P(W^c) = 1 \times 1/3 = 1/3$$

The $P(W \text{ and } C^c)$ is the numerator of the conditional probability that we need, and to calculate the denominator we have

$P(C^c) = P(W \text{ and } C^c) + P(W^c \text{ and } C^c) = 1/2 + 1/3 = 5/6$

These two calculations can be used to evaluate the probability that Julianne is a carrier given that she and Jason have one child who does not have cystic fibrosis, namely

$$P(W \mid C^c) = \frac{P(W \text{ and } C^c)}{P(C^c)} = 3/5$$

We see the information that the first child does not have cystic fibrosis reduces the probability that Julianne is a carrier.

CHAPTER 5

SAMPLING DISTRIBUTIONS

SECTION 5.1

OVERVIEW

One of the most common situations giving rise to a **count** X is the **binomial setting.** The four assumptions about how the count was produced are

- the number n of observations is fixed.
- the n observations are all independent.
- each observation falls into one of two categories called "success" and "failure."
- the probability of success p is the same for each observation.

When these assumptions are satisfied, the number of successes X has a **binomial distribution** with n trials and success probability p, denoted by $B(n, p)$. For smaller values of n, the probabilities for X can be found easily using statistical software. Table C in the text gives the probabilities for certain combinations of n and p, and there is also an exact formula. For large n, the **normal approximation** can be used.

For a large population containing a proportion p of successes, the binomial distribution is a good approximation to the number of successes in an SRS of size n, provided the population is at least 20 times larger than the sample. The mean and standard deviation for the binomial count X and the sample proportion $\hat{p} = X/n$ can be found using the formulas

$$\mu_X = np \qquad \mu_{\hat{p}} = p$$

$$\sigma_X = \sqrt{np(1-p)} \qquad \sigma_{\hat{p}} = \sqrt{\frac{p(1-p)}{n}}$$

When n is large the count X is approximately $N\left(np, \sqrt{np(1-p)}\right)$ and the proportion $\hat{p}$ is approximately $N\left(p, \sqrt{\frac{p(1-p)}{n}}\right)$. These approximations should work well when $np \geq 10$ and $n(1-p) \geq 10$. For the count X, the **continuity correction** improves the approximation, particularly when the values of n and p are closer to these cutoffs.

The exact **binomial probability formula** is given by

$$P(X = k) = \binom{n}{k} p^k (1-p)^{n-k}$$

where $k = 0, 1, 2, \ldots, n$ and $\binom{n}{k}$ is the **binomial coefficient**

$$\binom{n}{k} = \frac{n!}{k!(n-k)!}$$

and the factorial $n!$ is

$$n! = n \times (n-1) \times (n-2) \times \cdots \times 3 \times 2 \times 1$$

GUIDED SOLUTIONS

Exercise 5.11

KEY CONCEPTS: Binomial setting

Four assumptions need to be satisfied to ensure that the count X has a binomial distribution. The number of observations or trials is fixed in advance, each trial results in one of two outcomes, the trials are independent, and the probability of success is the same from trial to trial. In addition, for a large population with a proportion p of successes, we can use the binomial distribution as an approximation to the distribution of the count X of successes in an SRS of size n, provided the population is at least 20 times the size of the sample. For each setting below, if the binomial distribution applies give the values of n and p.

(a) Think about how this fits in the binomial setting. What is n? What are the two outcomes? This is an example of the binomial distribution in statistical sampling. You will need to describe p instead of giving a numerical value. Is the population large enough to use the binomial distribution?

(b) Think about how many observations or trials there are going to be.

(c) This is an example of the binomial distribution in statistical sampling. What constitutes a trial and are there only two outcomes? What would correspond to the success probability? What are the possible values of X?

Exercise 5.13

KEY CONCEPTS: Binomial probabilities, binomial tables

(a) Suppose we let X denote the number of errors caught. There are 10 word errors in the essay, and we are in the binomial setting with $n = 10$ trials. Letting "success" correspond to the students catching a word error, we have $p = 0.7$. The distribution of the number of errors caught is $B(10, 0.7)$.

Now, suppose Y denotes the number of errors missed. What are the values of n and p? What is the distribution of the number of errors missed?

(b) We want the probability that 4 or more errors are missed. Express this probability in terms of Y, the number of errors missed. Add the appropriate entries in Table C to compute this probability. Alternatively, you may wish to use statistical software to compute the probability.

Exercise 5.15

KEY CONCEPTS: Binomial distribution, mean and variance

(a) Suppose we let X denote the number of errors caught. The distribution of the number of errors caught is $B(10, 0.7)$. The mean of X is

$\mu_X = np =$

Now, suppose Y denotes the number of errors missed. The mean of Y is

$\mu_Y = np =$

What is the total of the number of errors caught plus the number of errors missed? Can you see why the means must add to 10?

(b) Suppose we let X denote the number of errors caught. The distribution of the number of errors caught is $B(10, 0.7)$. The standard deviation of X is

$\sigma_X = \sqrt{np(1-p)} =$

(c) Suppose $p = 0.9$. The standard deviation of X is

$\sigma_X = \sqrt{np(1-p)} =$

Now suppose $p = 0.99$. The standard deviation of X is

$\sigma_X = \sqrt{np(1-p)} =$

What happens to the standard deviation of a binomial distribution as the success probability gets closer to 1?

Exercise 5.17

KEY CONCEPTS: Binomial distribution, binomial tables

The number of misses has a binomial distribution with $n = 10$ and $p = 0.3$. You can use Table C to evaluate $P(X \geq m)$ for any value of m. The larger the value of m, the smaller this probability becomes. The problem asks you to find the smallest value of m so that this probability is less than 0.05.

Why might you consider m or more misses as evidence that a proofreader actually catches less than 70% of nonwords?

Exercise 5.19

KEY CONCEPTS: Binomial probabilities, mean of the binomial

(a) Let X denote the number of fives in a group of five digits from the table of random digits. The number of fives has a binomial distribution. What is n, the number of trials, and what is the numerical value of the success probability p? Fill in the values of n and p below.

$n =$

$p =$

Now write the event that a group of five digits from the table will contain at least one 5 in terms of X, and then use Table C to evaluate this probability.

(b) Now let X denote the number of fives in a line 40 digits long in the table of random digits. The distribution of X is again binomial. First fill in the values of n and p and then use these in the formula for the mean of X.

$n =$

$p =$

$\mu_X = np =$

Exercise 5.25

KEY CONCEPTS: Proportions, normal approximation for proportions

(a) Let X denote the number of students in the SRS of n = 200 who support a crackdown on underage drinking. In this case, X = 140. To compute the sample proportion use the formula

$$\text{sample proportion} = \hat{p} = X/n =$$

(b) If the proportion of all students on your campus who support a crackdown is p = 0.67, then X is a count having the B(200, 0.67) distribution. You need to compute a probability about $\hat{p}$, namely $P(\hat{p} \geq 0.70)$, because 0.70 is the sample proportion that resulted from the administration's sample. Since $\hat{p} \geq 0.70$ whenever $X \geq 140$, if you have statistical software which computes binomial probabilities, try to find the exact probability.

If you don't have access to statistical software that computes binomial probabilities, you'll need to use the normal approximation to the sampling distribution of $\hat{p}$ to approximate the probability. Since np = 200(0.67) = 133 and $n(1 - p)$ = 200(0.33) = 67 are both greater than or equal to 10, we can use the approximation. To use the normal approximation, the mean and standard deviation of $\hat{p}$ must be computed. Do this using the formulas given below.

$$\mu_{\hat{p}} = p =$$

$$\sigma_{\hat{p}} = \sqrt{\frac{p(1-p)}{n}} =$$

Computing $P(\hat{p} \geq 0.70)$ is a normal probability calculation like those in Section 3 of Chapter 1. You may want to review the material there to refresh your memory on how to do such calculations, as they form the basis of solving many of the exercises in this chapter. To compute the probability, we first standardize the number 0.62 (compute its z-score) by subtracting the mean $\mu_{\hat{p}}$ and dividing the result by $\sigma_{\hat{p}}$. Next, use Table A to determine the area to the right of this z-score. It may be helpful to draw a normal curve to help you visualize the desired area.

$P(\hat{p} \geq 0.70) =$

(c) You should base your comments on the probability calculated in (b). In particular, is this probability large enough that a value as large as 0.70 could plausibly arise by chance if the actual value is 0.67?

Exercise 5.31

KEY CONCEPTS: Normal approximation for counts

(a) The population proportion is

$$p = \frac{\text{number of blacks in the 2000 census}}{\text{total number of adults}}$$

Use the numbers given in the statement of the problem to compute p.

$p =$

(b) Let X denote the number of blacks in the sample. X is a count and if the sample is a random sample, X will have a binomial distribution. What are n and p in this case? Now use the formulas that express the mean of X in terms of n and p.

mean =

(c) This part can be done using the normal approximation for a count. First, check to see that np and $n(1 - p)$ are both greater than or equal to 10. Next, compute the standard deviation of X using the formula that expresses the standard deviation in terms of n and p.

standard deviation =

Now use the mean from (b), this standard deviation, and the normal approximation to evaluate

$P(X \leq 100) =$

COMPLETE SOLUTIONS

Exercise 5.11

(a) This is an example of choosing an SRS from a population. The number of trials, 200, is fixed and the size of the population of college students should be at least 20 times the size of the sample. The success probability p is the proportion of all college students who would reply that they are irritable in the morning, but this proportion is not known. The binomial distribution with $n = 200$ and p as the proportion of *all* college students who would reply that they are irritable in the morning should be a good probability model for the number of college students among the 200 who reply that they are usually irritable in the morning.

(b) Although each toss of the coin is a success (Head) or a failure (Tail), we are not counting the number of successes in a fixed number of tosses. The number of observations (tosses) is random. The assumption of a fixed number of observations is violated.

(c) The number of trials, 500, is fixed and each call either succeeds in talking with a live person or it doesn't. The number of possible phone numbers in New York City is at least 20 times the sample size and the probability of any call succeeding is given as 1/12, the proportion in the population. The binomial distribution with n = 500 and p = 1/12 should be a good probability model for the number of calls that reach a live person. The values of X are any integer between 0 and 500.

Exercise 5.13

(a) The distribution of the number of errors caught is given in the Guided Solutions. If Y denotes the number of errors missed, then n = 10 and $p = 1 - 0.7 = 0.3$, where p is now the probability of missing a given error. The distribution of the number of errors missed is $B(10, 0.3)$.

(b) Letting Y denote the number of errors missed, recall that the distribution is $B(10, 0.3)$.

$$P(Y \geq 4) = P(Y = 4) + P(Y = 5) + P(Y = 6) + \cdots + P(Y = 10)$$

Using Table C to find each of the probabilities on the right-hand side of this equation we find

$$P(Y \geq 4) = 0.2001 + 0.1029 + 0.0368 + 0.0090 + 0.0014 + 0.0001 + 0.0000 = 0.3503,$$

where the probability corresponding to $Y = 10$ is taken to be zero (Software gives $P(Y \geq 4) = 0.3504$).

Exercise 5.15

(a) If X denotes the number of errors caught, the mean of X is $\mu_X = np = 10(0.7) = 7$. Now, suppose Y denotes the number of errors missed. The mean of Y is $\mu_Y = np = 10(0.3) = 3$. We see that these means add to 10. In any experiment, the total of the number of errors caught plus the number of errors missed must *always* be 10, so 10 must be the mean of this total.

(b) If X is the number of errors caught, the standard deviation of X is

$$\sigma_X = \sqrt{np(1-p)} = \sqrt{10(0.7)(0.3)} = 1.45$$

(c) Suppose $p = 0.9$. The standard deviation of X is

$$\sigma_X = \sqrt{np(1-p)} = \sqrt{10(0.9)(0.1)} = 0.95$$

Now suppose $p = 0.99$. The standard deviation of X is

$$\sigma_X = \sqrt{np(1-p)} = \sqrt{10(0.99)(0.01)} = 0.31$$

As the success probability gets closer to 1, the standard deviation gets closer to 0. When the success probability is very close to 1 there is very little variability in the outcome as the proofreader will tend to catch all 10 errors almost every time.

Exercise 5.17

Going to Table C with $n = 10$ and $p = 0.3$ we find

$$P(X \geq 6) = + 0.0368 + 0.0090 + 0.0014 + 0.0001 + 0.0000 = 0.0473,$$

where the probability corresponding to $X = 10$ is taken to be zero. From Table C we see that $P(X = 5) = 0.1029$, so the value of m must be 6. Since the chance of missing 6 or more errors is so small, missing 6 or more errors could be considered evidence that the proofreader actually is catching under 70% of the errors, or equivalently the proofreader is missing more than 30% of the errors.

Exercise 5.19

KEY CONCEPTS: Binomial probabilities, mean of the binomial, normal approximation for counts

(a) Let X denote the number of fives in a group of five digits from the table of random digits. Each digit is either a 5 or it isn't so there are two outcomes. The digits are independent of each other and the number of trials is $n = 5$. Letting a digit of 5 correspond to a success, the success probability is $p = 0.10$, the chance of any digit being a five.

The event that a group of five digits from the table will contain at least one 5 can be written in terms of X as $P(X \geq 1)$. Using Table C to evaluate this probability we have

$P(X \geq 1) = 1 - P(X = 0) = 1 - 0.5905 = 0.4095.$

(b) Now let X denote the number of fives in a line 40 digits long in the table of random digits. The distribution of X is again binomial, but with $n = 40$ and $p = 0.10$. The formula for mean of X gives

$$\mu_X = np = 40 \times 0.10 = 4$$

Exercise 5.25

(a) Sample proportion = $\hat{p} = X/n = 140/200 = 0.70$

(b) To use the normal approximation we compute

$$\text{mean} = \mu_{\hat{p}} = p = 0.67$$

$$\text{standard deviation} = \sigma_{\hat{p}} = \sqrt{\frac{p(1-p)}{n}} = \sqrt{\frac{0.67 \times 0.33}{200}} = \sqrt{0.001106} = 0.0333.$$

From this, we compute the z-score of 0.70 to be

$$z\text{-score} = \frac{0.70 - 0.67}{0.0333} = 0.90$$

Using Table A we find

$$P(\hat{p} \geq 0.70) = P(Z \geq 0.90) = 1 - 0.8159 = 0.1841$$

(*Note*: Using statistical software we find the exact probability that $\hat{p} \geq 0.70$, or equivalently that

$X \geq 140$, to be 0.2049. This differs somewhat from the value obtained using normal approximation).

(c) It is true that the sample proportion of 0.70 is larger than the national value of 0.67. However, our probability calculation shows that a sample value of 0.70 (or larger) has probability 0.1814 of occurring simply by chance if the actual proportion on campus is 0.67. In most scientific journals this probability would be considered too large to allow one to assert with any degree of confidence that the survey supports the statement that "support for a crackdown is higher on our campus than nationally."

Exercise 5.31

(a) $p = \dfrac{23{,}772{,}494}{209{,}128{,}094} = 0.1137$

(b) X has a binomial distribution with $n = 1200$ and $p = 0.1137$. The mean of X is

$$\text{mean} = np = 1200(0.1137) = 136.44$$

(c) The normal approximation can be used since both $np = 136.44$ and $n(1 - p) = 1063.56$ are greater than 10. We compute

$$\text{standard deviation} = \sqrt{np(1-p)} = \sqrt{(1200)(0.1137)(0.8863)} = 10.997$$

Using the mean in (b) and this standard deviation gives the approximation

$$P(X \leq 100) = P\left(\frac{X - 136.44}{10.997} \leq \frac{100 - 136.44}{10.997}\right) = P(Z \leq -3.31) = 0.0005$$

SECTION 5.2

OVERVIEW

This section examines properties of the **sample mean** $\bar{x}$. If we select an SRS of size n from a large population with mean μ and standard deviation σ, the sample mean $\bar{x}$ has a sampling distribution with

$$\text{mean} = \mu_{\bar{x}} = \mu$$

and

$$\text{standard deviation} = \sigma_{\bar{x}} = \frac{\sigma}{\sqrt{n}}$$

This implies that the sample mean is an unbiased estimator of the population mean and is less variable than a single observation.

Linear combinations (such as sums or means) of independent normal random variables have normal distributions. In particular, if the population has a normal distribution, the sampling distribution of $\bar{x}$ is normal. Even if the population does not have a normal distribution, for large sample sizes the sampling distribution of $\bar{x}$ computed from an SRS is approximately normal. In particular, the **central limit**

theorem states that for large n, the sampling distribution of $\bar{x}$ computed from an SRS is approximately $N\left(\mu, \frac{\sigma}{\sqrt{n}}\right)$ for any population with mean μ and finite standard deviation σ.

GUIDED SOLUTIONS

EXERCISE 5.50

KEY CONCEPTS: The sampling distribution of the sample mean, normal probability calculations

(a) We take X to be the (numeric) grade of a randomly chosen student. The probability distribution of X is seen to be

Grade (X)	4	3	2	1	0
Probability	0.31	0.40	0.20	0.04	0.05

Refer to Section 4.4 for the formulas for the mean and standard deviation of a discrete random variable. Use these to compute the following.

μ = mean of X =

σ^2 = variance of X =

σ = standard deviation of X =

(b) Recall that if we select an SRS of size n from a large population with mean μ and standard deviation σ, the sample mean $\bar{x}$ has a sampling distribution with

mean of $\bar{x} = \mu_{\bar{x}} =$

and

standard deviation of $\bar{x} = \sigma_{\bar{x}} = \frac{\sigma}{\sqrt{n}}$

You have computed the values of μ and σ in part (a). In this case, what is n, the sample size? Use these to fill in the values for the mean and standard deviation of $\bar{x}$ above.

(c) You can compute the probability $P(X \geq 3)$ that a randomly chosen English 210 student gets a grade of B or better directly using the probability distribution in part (a). This is just like the calculations in Section 4.3.

$P(X \geq 3) =$

To compute the approximate probability $P(\overline{x} \geq 3)$ that the grade point average for 50 randomly chosen English 210 students is B or better, we use the central limit theorem, which states that for large n, the sampling distribution of $\overline{x}$ computed from an SRS is approximately $N\left(\mu, \frac{\sigma}{\sqrt{n}}\right)$ for any population with mean μ and finite standard deviation σ. Thus, calculating $P(\overline{x} \geq 3)$ is just a normal probability calculation, like those we did in Section 3 of Chapter 1. You may wish to review the material there to refresh your memory as to how to do such calculations. We want the probability that the sample mean $\overline{x}$ is 3 or higher. To compute this probability, we standardize the number 3 (compute its z-score) by subtracting the mean $\mu_{\overline{x}}$ and dividing the result by $\sigma_{\overline{x}}$. You computed $\mu_{\overline{x}}$ and $\sigma_{\overline{x}}$ in (b). Next, use Table A to determine the area to the right of this z-score under a standard normal curve. You may wish to draw a picture of the standard normal curve to help you visualize the desired area.

$P(\overline{x} \geq 3) =$

EXERCISE 5.51

KEY CONCEPTS: Normal populations, the sampling distribution of the sample mean

(a) We are told that the distribution of Sheila's measured glucose level one hour after ingesting the sugary drink is approximately normal with mean $\mu = 125$ mg/dl and standard deviation $\sigma = 10$ mg/dl. Based on a single measurement, Sheila will be diagnosed as having gestational diabetes if this measurement exceeds 140 mg/dl. This is the type of problem we did in Section 3 of Chapter 1. For a normal population with $\mu = 125$ mg/dl and $\sigma = 10$ mg/dl, you must find the area to the right of 140 mg/dl. You may wish to review the material in Section 1.3 to refresh your memory as to how to do such calculations.

(b) We are told that the distribution of Sheila's measured glucose level one hour after ingesting the sugary drink is approximately normal with mean μ = 125 mg/dl and standard deviation σ = 10 mg/dl. Measurements are made on 3 separate days and averaged to obtain the mean $\bar{x}$. We will assume these three measurements are independent. Use this information to calculate the mean and standard deviation of the sample mean below.

mean $= \mu_{\bar{x}} = \mu =$

standard deviation $= \sigma_{\bar{x}} = \dfrac{\sigma}{\sqrt{n}} =$

The average of three independent normal random variables has a normal distribution. You have just calculated the mean and standard deviation of this distribution. Now use these values and the normal probability calculations in Section 3 of Chapter 1 to find the probability that this average exceeds 140. That is, compute

$P(\bar{x} \geq 140)$

EXERCISE 5.53

KEY CONCEPTS: Sampling distribution of the sample mean, normal probability calculations

We are told that the distribution of Sheila's measured glucose level one hour after ingesting the sugary drink is approximately normal with mean $\mu = 125$ and standard deviation $\sigma = 10$. To find L, you need to first determine the sampling distribution of the mean score $\bar{x}$ of the $n = 3$ measurements. This sampling distribution was obtained in part (b) of Exercise 5.53.

To complete the problem, you need to find L such that the probability of $\bar{x}$ being above L is only 0.05. We did this sort of problem in Section 3 of Chapter 1. First, refer to Table A to find the value z such that the area to the *left* of z under a standard normal curve is 0.95, since we are told that the area to the right must be 0.05. What is this z-value?

$z =$

This value z is the z-score of L. This means $z = (L - \mu_{\bar{x}})/\sigma_{\bar{x}}$, where $\mu_{\bar{x}}$ and $\sigma_{\bar{x}}$ are the mean and standard deviation, respectively, of the sampling distribution of $\bar{x}$. Solve this equation for L.

EXERCISE 5.59

KEY CONCEPTS: Linear combinations of independent random variables

(a) We know that the breaking strengths of untreated specimens are normally distributed. Thus, the mean $\bar{x}$ of 6 untreated specimens is normally distributed with

mean $= \mu_{\bar{x}} = \mu =$

standard deviation $= \sigma_{\bar{x}} = \dfrac{\sigma}{\sqrt{n}} =$

Use the mean and standard deviation to calculate the probability that the mean breaking strength of the 6 untreated specimens exceeds 50 pounds, that is,

$P(\bar{x} \geq 50)$

(b) Letting $\bar{x}$ and $\bar{y}$ be the sample means of the untreated and the treated specimens, respectively, we need to obtain some results about the sampling distribution of $\bar{x} - \bar{y}$ in order to complete this problem. To do this, we need to first recall several facts from Section 4 of Chapter 4. For a random variable X

$$\mu_{a+bX} = a + b\,\mu_X$$

Using $a = 0$ and $b = -1$, this implies that $\mu_{-X} = -\mu_X$. Second, recall that for random variables X and Y

$$\mu_{X+Y} = \mu_X + \mu_Y$$

Replacing X by the random variable $-X$ and using $\mu_{-X} = -\mu_X$ we have

$$\mu_{X-Y} = \mu_X - \mu_Y$$

Finally, recall that if X and Y are independent random variables, then

$$\sigma^2_{X-Y} = \sigma^2_X + \sigma^2_Y$$

Getting back to our problem, letting $\bar{x}$ and $\bar{y}$ be the sample means of the untreated and the treated specimens, respectively, we can apply the previous results to $\bar{x}$ and $\bar{y}$ to show

$$\mu_{\bar{x}-\bar{y}} = \mu_{\bar{x}} - \mu_{\bar{y}}$$

and

$$\sigma^2_{\bar{x}-\bar{y}} = \sigma^2_{\bar{x}} + \sigma^2_{\bar{y}}$$

We take the square root of $\sigma^2_{\bar{x}-\bar{y}}$ to get the standard deviation of $\bar{x} - \bar{y}$. The problem asks you to calculate the probability that the mean breaking strength of the 6 untreated specimens is at least 25 pounds greater than the mean strength of the 6 treated specimens, or $P(\bar{x}-\bar{y}>25)$. In order to determine this probability, we need to use the previous results to compute the mean and standard deviation of the sampling distribution of $\bar{x} - \bar{y}$. Complete the following calculations. You need to be careful when calculating the variances $\sigma^2_{\bar{x}}$ and $\sigma^2_{\bar{y}}$. The formulas given in the Overview are for the standard deviations $\sigma_{\bar{x}}$ and $\sigma_{\bar{y}}$ and must be squared to get the variances.

$\mu_{\bar{x}} =$ $\sigma^2_{\bar{x}} =$

$\mu_{\bar{y}} =$ $\sigma^2_{\bar{y}} =$

$\mu_{\bar{x}-\bar{y}} = \mu_{\bar{x}} - \mu_{\bar{y}} =$ $\sigma^2_{\bar{x}-\bar{y}} = \sigma^2_{\bar{x}} + \sigma^2_{\bar{y}} =$

$\sigma_{\bar{x}-\bar{y}} =$

Now that you have the mean and standard deviation of $\bar{x} - \bar{y}$, it is a straightforward normal calculation to find

$P(\bar{x}-\bar{y}>25) =$

Complete Solutions

EXERCISE 5.50

(a) μ = mean of X = $4 \times 0.31 + 3 \times 0.40 + 2 \times 0.20 + 1 \times 0.04 + 0 \times 0.05 = 2.88$

To compute the standard deviation we first compute the variance

$$\sigma^2 = (4-2.88)^2 \times 0.31 + (3-2.88)^2 \times 0.40 + (2-2.88)^2 \times 0.20 + (1-2.88)^2 \times 0.04$$
$$+ (0-2.88)^2 \times 0.05 = 1.1056$$

and so

σ = standard deviation of X = $\sqrt{1.1056} = 1.0515$

(b) Using the results in part (a) and the fact that the sample size is $n = 50$, we obtain

$$\text{mean of } \bar{x} = \mu_{\bar{x}} = \mu = 2.88$$

and

$$\text{standard deviation of } \bar{x} = \sigma_{\bar{x}} = \frac{\sigma}{\sqrt{n}} = \frac{1.0515}{\sqrt{50}} = 0.1487$$

(c) Using the probability distribution for X, we have

$$P(X \geq 3) = P(X = 3) + P(X = 4) = 0.40 + 0.31 = 0.71$$

Now, using the results of (b), the sampling distribution of $\bar{x}$ is approximately $N(2.88, 0.1487)$. Thus,

$$P(\bar{x} \geq 3) = P\left(\frac{\bar{x} - 2.88}{0.1487} \geq \frac{3 - 2.88}{0.1487}\right) = P(Z \geq 0.81) = 1 - P(Z \leq 0.81) = 1 - 0.7910 = 0.2090.$$

EXERCISE 5.51

(a) For a normal population with $\mu = 125$ mg/dl and $\sigma = 10$ mg/dl, the area to the right of 140 mg/dl is found as

$$P(X > 140) = P\left(\frac{X - 125}{10} > \frac{140 - 125}{10}\right) = P(Z > 1.5) = 1 - P(Z < 1.5) = 1 - 0.9332 = 0.0668$$

(b) The sampling distribution of the average of three glucose measurements has

$$\text{mean } = \mu_{\bar{X}} = \mu = 125$$

and

$$\text{standard deviation } = \sigma_{\bar{X}} = \frac{\sigma}{\sqrt{n}} = \frac{10}{\sqrt{3}} = 5.773$$

The sampling distribution of $\bar{x}$ is normal and the probability that Sheila is diagnosed with gestational diabetes on the basis of the average of three measurements is

$$P(\bar{x} > 140) = P\left(\frac{\bar{x} - 125}{5.773} > \frac{140 - 125}{5.773}\right) = P(Z > 2.60) = 1 - P(Z < 2.60) = 1 - 0.9953 = 0.0047$$

EXERCISE 5.53

Using part (b) of Exercise 5.51, the sampling distribution of the mean score $\bar{x}$ of the $n = 3$ glucose measurements is normal with mean

$$\mu_{\bar{x}} = \mu = 125$$

and standard deviation

$$\sigma_{\bar{x}} = \frac{\sigma}{\sqrt{n}} = 5.773$$

Next, we note from Table A the value of z such that the area to the left of it under a standard normal curve is 0.95 is $z = 1.65$. Thus,

$$1.65 = (L - 125)/5.773$$

Solving for L gives

$$L = (1.65)(5.773) + 125 = 134.53.$$

EXERCISE 5.59

(a) The mean $\bar{x}$ of 6 untreated specimens is normally distributed with

$$\text{mean } = \mu_{\bar{x}} = \mu = 57$$

and

$$\text{standard deviation } = \sigma_{\bar{x}} = \frac{\sigma}{\sqrt{n}} = \frac{2.2}{\sqrt{6}} = 0.898$$

Using standard normal calculations as in Section 3 of Chapter 1,

$$P(\bar{x} > 50) = P\left(\frac{\bar{x} - 57}{0.898} > \frac{50 - 57}{0.898}\right) = P(Z > -7.80) = 1 - P(Z < -7.80) \approx 1$$

(b) Remembering to square the standard deviations because the formulas are in terms of variances, and letting x correspond to the untreated group and y to the treated group, we have for the untreated group

$$\mu_{\bar{x}} = \mu_x = 57$$

$$\sigma_{\bar{x}}^2 = \frac{\sigma_x^2}{n} = \frac{(2.2)^2}{6} = 0.8067$$

A similar calculation for the treated specimens gives

$$\mu_{\bar{y}} = \mu_y = 30$$

$$\sigma_{\bar{y}}^2 = \frac{\sigma_y^2}{n} = \frac{(1.6)^2}{6} = 0.4267$$

Using these results for the distribution of $\bar{x} - \bar{y}$ gives

$$\mu_{\bar{x}-\bar{y}} = \mu_{\bar{x}} - \mu_{\bar{y}} = 57 - 30 = 27$$

$$\sigma^2_{\bar{x}-\bar{y}} = \sigma^2_{\bar{x}} + \sigma^2_{\bar{y}} = 0.8067 + 0.4267 = 1.2334$$

and

$$\sigma_{\bar{x}-\bar{y}} = \sqrt{1.2334} = 1.1106$$

Now that you have the mean and standard deviation of $\bar{x} - \bar{y}$, it is a straightforward normal calculation to find

$$P(\bar{x}-\bar{y} > 25) = P\left(\frac{\bar{x}-\bar{y}-27}{1.1106} > \frac{25-27}{1.1106}\right) = P(Z > -1.80) = 1 - P(Z < -1.80) = 1 - 0.0359 = 0.9641$$

CHAPTER 6

INTRODUCTION TO INFERENCE

SECTION 6.1

OVERVIEW

A **confidence interval** provides an estimate of an unknown parameter of a population or process along with an indication of how accurate this estimate is and how **confident** we are that the interval is correct. Confidence intervals have two parts. One is an interval computed from our data. This interval typically has the form

estimate ± margin of error

The other part is the **confidence level**, which states the probability that the *method* used to construct the interval will give a correct answer. For example, if you use a 95% confidence interval repeatedly, in the long run 95% of the intervals you construct will contain the correct parameter value. Of course, when you apply the method only once, you do not know if your interval gives a correct value or not. "Confidence" refers to the probability that the method gives a correct answer in repeated use, not the correctness of any particular interval we compute from data.

Suppose we wish to estimate the unknown mean μ of a normal population with known standard deviation σ based on an SRS of size n. A level C confidence interval for μ is

$$\bar{x} \pm z^* \frac{\sigma}{\sqrt{n}}$$

where z^* is such that the probability is C that a standard normal random variable lies between $-z^*$ and z^* and is obtained from the bottom row in Table D.

The margin of error $z^* \frac{\sigma}{\sqrt{n}}$ of a confidence interval decreases when any of the following occur:

- the confidence level C decreases
- the sample size n increases
- the population standard deviation σ decreases.

The sample size needed to obtain a confidence interval for a normal mean of the form

estimate ± margin of error

with a specified margin of error m is

$$n = \left(\frac{z^* \sigma}{m}\right)^2$$

where z^* is the critical point for the desired level of confidence. Many times the n you will find will not be an integer. If it is not, round up to the next larger integer.

The formula for any specific confidence interval is a recipe that is correct under specific conditions. The most important conditions concern the methods used to produce the data. Many methods (including those discussed in this section) assume that our data were collected by random sampling. Other conditions, such as the actual distribution of the population, are also important.

GUIDED SOLUTIONS

Exercise 6.17

KEY CONCEPTS: Margin of error, confidence intervals

The margin of error of a level C confidence interval is $z^* \frac{\sigma}{\sqrt{n}}$, where z^* is such that the probability is C that a standard normal random variable lies between $-z^*$ and z^* and is obtained from the bottom row in Table D. To do this exercise, you must identify

C = the level of confidence required =

z^* = the probability is C that a standard normal random variable lies between $-z^*$ and z^* =

σ = population standard deviation =

n = the sample size used =

Determine the above values and then compute the margin of error

$$m = z^* \frac{\sigma}{\sqrt{n}} =$$

The confidence interval is of the form estimate ± margin of error. Give the 95% confidence interval below.

Exercise 6.19

KEY CONCEPTS: Margin of error, confidence intervals

The margin of error $z^* \frac{\sigma}{\sqrt{n}}$ of a confidence interval decreases when any of the following occur:

- the confidence level C decreases
- the sample size n increases
- the population standard deviation σ decreases

For both samples we are computing 95% confidence intervals based on a sample of 100 sophomore students. What is the important difference between the population of the amount spent on textbooks in the fall by all sophomore students at your university and the population of the amount spent on textbooks in the fall by all sophomore students in your major that affects the margin of error? Which sample would have the smaller margin of error?

Exercise 6.25

KEY CONCEPTS: The sampling distribution of $\bar{x}$, confidence intervals

(a) We are told that the $n = 200$ women are a random sample, and we know that the sampling distribution of $\bar{x}$ has standard deviation $\sigma_{\bar{x}} = \frac{\sigma}{\sqrt{n}}$. Use the values of σ and n to compute $\sigma_{\bar{x}}$.

$\sigma_{\bar{x}} =$

(b) Recall from Section 3 of Chapter 1 the 68–95–99.7 rule, which says that 68% of the area under a normal curve lies within one standard deviation of the mean, 95% within two standard deviations of the mean, and 99.7% within three standard deviations of the mean. Now you should be able to fill in the blank.

(c) Use the value of $\sigma_{\bar{x}}$ that you computed in part (a) and your answer to part (b) to fill in the blank.

Exercise 6.29

KEY CONCEPTS: The sample size required to obtain a confidence interval of specified margin of error

The standard deviation of the biomarker measurements is known to be 6.5 U/l. The smallest value of n that will yield a 95% confidence interval with a margin of error = m must satisfy

$$n = \left(\frac{z^{*}\sigma}{m}\right)^2$$

where z^* is the critical point for the desired level of confidence and σ is given above. Identify z^*, σ, and m in this case and use the above formula to compute n. Remember to round your answer up to the nearest integer.

Exercise 6.33

KEY CONCEPTS: Confidence levels for several confidence intervals simultaneously, binomial probability calculations

We know the following:

- we are interested in a fixed number of intervals (five, to be precise)
- the five intervals are all independent
- either an interval contains the true mean (success) or it does not (failure)
- the probability that any particular interval will contain the true mean is 0.95

Let X be the number of intervals (out of the five) that contain the true mean (i.e., the number of successes). X should remind you of a special type of random variable whose probability distribution we have studied previously (*Hint*: Look at Section 1 of Chapter 5). How do you calculate probabilities for X?

(a) We want the probability of five successes. What is this probability?

(b) We want the probability of at least four successes. What is this probability?

Complete Solutions

Exercise 6.17

From the statement of the problem we see

$$C = \text{the level of confidence required} = 0.95$$

Hence,

z^* = the probability is 0.95 that a standard normal random variable lies between $-z^*$ and z^*

= 1.96 (see Table D)

We also see that

σ = population standard deviation = 6.5

n = the sample size used = 31

Thus, the margin of error m is

$$z^* \frac{\sigma}{\sqrt{n}} = 1.96 \frac{6.5}{\sqrt{31}} = 2.288 \text{ U/l}$$

and the 95% confidence interval is

$$13.2 \pm 2.288 \text{ U/l} \quad \text{or} \quad 10.91 \text{ U/l to } 15.49 \text{ U/l}$$

Exercise 6.19

The amount spent on textbooks by sophomores in your major at the university should be less variable than the amount spent among all sophomores, as those in your major will be taking many of the same courses with the same cost for textbooks. Thus, the population standard deviation σ should be smaller for the amount spent on textbooks restricted to sophomores in your major, and the margin of error will be smaller for the 95% confidence interval based on a sample of 100 sophomores in your major.

Exercise 6.25

(a) We are given that $\sigma = 31$ calories and $n = 200$. Hence, $\sigma_{\bar{x}} = \frac{\sigma}{\sqrt{n}} = \frac{31}{\sqrt{200}} = 2.192$.

(b) The probability is 0.95 that $\bar{x}$ is within $2 \times \sigma_{\bar{x}} = 4.384$ calories of the population mean μ.

(c) $2 \times \sigma_{\bar{x}} = 2 \times 2.192 = 4.384$ calories, so about 95% of all samples will capture the true mean number of calories consumed in the interval $\bar{x} \pm 4.384$ calories.

Exercise 6.29

From the statement of the problem we see

$$C = \text{the level of confidence required} = 0.95$$

Hence,

$z^* =$ the probability is 0.95 that a standard normal random variable lies between $-z^*$ and z^*

$= 1.96$ (see Table D)

We are given $\sigma =$ population standard deviation $= 6.5$ U/l and are told we want $m = 2$ U/l. Substituting these values in the formula for n we get

$$n = \left(\frac{z^*\sigma}{m}\right)^2 = \left(\frac{(1.96)(6.5)}{2}\right)^2 = (6.37)^2 = 40.5769$$

Rounding up, we see that the smallest value of n that will accomplish our goal is $n = 41$.

Exercise 6.33

Referring to Section 1 of Chapter 5 we see that X has a binomial $B(n = 5, p = 0.95)$ distribution. We use this distribution to calculate the required probabilities. Table C in your text gives binomial probabilities.

(a) Here we want the probability that $X = 5$. Table C only gives binomial probabilities for $p \leq 0.50$. Thus, to use Table C we have to rewrite the desired probability in terms of the number of failures (which has $p = 0.05$, a value that is given in Table C). We have (look in the portion of Table C with $n = 5$ and $p = 0.05$)

$$P(X = 5) = P(\text{number of failures} = 0) = 0.7738$$

(b) Again, rewriting the desired probability (at least 4 successes) in terms of number of failures we have

$$P(X \geq 4) = P(\text{number of failures} \leq 1) = P(0 \text{ failures}) + P(1 \text{ failure})$$

$$= 0.7738 + 0.2036 = 0.9774$$

SECTION 6.2

OVERVIEW

Tests of significance and confidence intervals are the two most widely used types of formal statistical inference. A test of significance is done to assess the evidence against the **null hypothesis** H_0 in favor of an **alternative hypothesis** H_a. Typically, the alternative hypothesis is the effect that the researcher is trying to demonstrate, and the null hypothesis is a statement that the effect is not present. The alternative hypothesis can be either **one-** or **two-sided**.

Tests are usually carried out by first computing a **test statistic**. The test statistic is used to compute a ***P*-value,** which is the probability of getting a test statistic at least as extreme as the one observed, where the probability is computed assuming that the null hypothesis is true. The *P*-value provides a measure of how incompatible our data is with the null hypothesis, or how unusual it would be to get data like ours if the null hypothesis were true. Since small *P*-values indicate data that are unusual or difficult to explain under the null hypothesis, we typically reject the null hypothesis in these cases. The alternative hypothesis provides a better explanation for our data.

Significance tests of the null hypothesis H_0: $\mu = \mu_0$ with either a one- or two-sided alternative are based on the test statistic

$$z = \frac{\overline{x} - \mu_0}{\sigma / \sqrt{n}}$$

The use of this test statistic assumes that we have an SRS from a normal population with known standard deviation σ. When the sample size is large, the assumption of normality is less critical because the sampling distribution of $\overline{x}$ is approximately normal. *P*-values for the test based on z are computed using Table A.

When the *P*-value is below a specified value α, we say the results are statistically significant at level α, or we reject the null hypothesis at level α. Tests can be carried out at a fixed significance level by obtaining the appropriate critical value z^* from the bottom row in Table D.

GUIDED SOLUTIONS

Exercise 6.53

KEY CONCEPTS: Null and alternative hypotheses

Typically, H_0 is of the form

$$H_0\text{: } \mu = \text{constant}$$

and H_a is of the form

$$H_a\text{: } \mu \neq \text{constant (two-sided alternative)}$$

or

$$H_a\text{: } \mu > \text{constant (one-sided alternative)}$$

or

$$H_a\text{: } \mu < \text{constant (one-sided alternative)}$$

Remember that in many instances (especially with one-sided alternatives) it is easier to begin with H_a, the effect that we are concerned about, and then to set up H_0 as the statement that the effect is absent.

In each example, think carefully about whether H_a should be one-sided or two-sided.

(a) We are interested in determining whether or not the population of students who pass the placement test have a mean score on a standard listening test of 26. That is, does the mean score of students in the courses apply to those who have taken the placement test? It will not apply if the mean differs from 26. Is H_a one- or two-sided? What is H_a? What is H_0?

(b) Why is the music being used? What do the researchers hope to show by playing the rap music? Use this to set up H_a and H_0. Is the alternative one- or two-sided?

(c) What are you concerned about in this example? Set up H_0 and H_a.

Exercise 6.59

KEY CONCEPTS: Relationship between two-sided tests and confidence intervals

(a) We are interested in determining whether or not the mean μ is 24 at the 10% significance level. State the null and alternative hypotheses.

H_0: H_a:

A level α two-sided significance test rejects a hypothesis H_0: $\mu = \mu_0$ when μ_0 falls outside a level $1 - \alpha$ confidence interval for μ. In this exercise, we are told that a 90% confidence interval for μ is (23, 28). Thus, α corresponds to 0.10 in this case and we can use the confidence interval to conduct a significance test at the 10% level. What is the value of μ_0 and is it outside the 90% confidence interval? Do we reject at the 10% significance level?

(b) What is μ_0 in this case? Can we reject the null hypothesis at the 10% significance level?

Exercise 6.65

KEY CONCEPTS: Interpreting significance levels

Write your explanation in the space. Refer to the Section 6.2 Overview in this Study Guide or Section 6.2 in the text if you need a hint.

Exercise 6.71

KEY CONCEPTS: Null and alternative hypotheses, carrying out a significance test about a mean

(a) What do the researchers hope to show? This will be the alternative. Write down the null and alternative hypotheses.

(b) The first step in carrying out the test is to compute the test statistic $z = \dfrac{\overline{x} - \mu_0}{\sigma / \sqrt{n}}$, which measures how far the sample mean is from the hypothesized value μ_0. To find the numerical value of z, you first need to determine μ_0, σ, and n from the problem, and then you need to compute $\overline{x}$, the mean difference between the computer's and the driver's calculations from a random sample of 20 records.

$z =$

Once the value of the test statistic z has been determined, the *P*-value can be computed. The *P*-value is the probability that the test statistic takes a value at least as extreme as the one observed. In the space provided, write this down as a probability in terms of Z, the standard normal random variable, and then use Table A to find this probability.

P-value =

If you are having trouble doing this directly from the meaning of the *P*-value, refer to the rules for computing *P*-values given in Section 6.2 in the text and try to understand the rationale behind them.
Now try to interpret your result in plain language.

Exercise 6.73

KEY CONCEPTS: Testing hypotheses at a fixed significance level

The value of the test statistic is given as $z = 2.36$. When carrying out the test at a fixed significance level, you can first compute the *P*-value and then reject the null hypothesis if the *P*-value is smaller than the significance level given. However, it is more direct to look up the critical value z^* in Table D and compare the value of the test statistic directly to the critical value.

(a) The advocacy group wants to see if the mean nicotine content is higher than 1.4. Write down the null and alternative hypotheses.

(b) The value of the test statistic is given as $z = 2.36$. When carrying out the test at a fixed significance level, you can first compute the *P*-value and then reject the null hypothesis if the *P*-value is smaller than the significance level given. However, it is more direct to look up the critical value z^* in Table D and compare the value of the test statistic directly to the critical value.

The alternative is $\mu > 1.4$, so the *P*-value is $P(Z > 2.36)$, which represents the probability of a test statistic at least as extreme as the one observed. Compare the *P*-value to 0.05. What do you conclude?

Using Table D, we look up the critical value corresponding to the tail probability 0.05 (it does not need to be doubled since the test is one-sided) and find the critical value $z^* = 1.645$. We will reject if the computed value of the test statistic exceeds this critical value. What do you conclude? When the significance level is fixed, it is easier to use Table D directly.

(c) Follow the procedure in (a), but with significance level 1%.

Exercise 6.79

KEY CONCEPTS: Calculating *P*-values

We are doing a two-sided test and we know from Table D that there is a 99.5% chance that z is between ± 2.807. Thus, the two-sided test at the .5% level rejects whenever $z < -2.807$ or $z > 2.807$. When does the .1% level test reject? Use these two rejection regions to give a value of z that will give a result that is significant at the 0.5% level but not at the 0.1% level.

COMPLETE SOLUTIONS

Exercise 6.53

(a) We are interested in determining whether the mean score on a standard listening test for those students who pass a placement test is 26 or not. Thus, the alternative hypothesis is two-sided and of the form H_a: $\mu \neq 26$. The null hypothesis is H_0: $\mu = 26$. The actual measurement, or the response, on each student will be their score on the listening test.

(b) The researchers believe that the music will cause the mice to complete the maze faster. An effect is present if the mean time μ to complete the maze is less than 20 seconds (the time to complete the maze with no stimulus). Hence, the alternative hypothesis is H_a: $\mu < 20$. The null hypothesis is thus H_0: $\mu = 20$ (one could also express the null hypothesis as H_0: $\mu \geq 20$).

(c) We are interested in determining whether the mean square footage of the population of one-bedroom apartments in a new student housing development is less than the advertised value of 460 square feet. Thus, the alternative hypothesis is one-sided and of the form H_a: $\mu < 460$. The null hypothesis is H_0: $\mu =$ 460.

Exercise 6.59

(a) The hypotheses are

$$H_0: \mu = 24 \qquad\qquad H_a: \mu \neq 24$$

In this case $\mu_0 = 24$ doesn't fall outside the level 90% confidence interval for μ, which is (23, 28). So we fail to reject H_0 at significance level 10%.

(b) Since $\mu_0 = 30$ corresponds to the null hypothesis H_0: $\mu = 30$, and the value 30 falls outside the 90% confidence interval for μ, we reject H_0 at significance level 10%.

Exercise 6.65

We are not told whether the average score on the science test for eighth graders throughout the nation in 2005 was higher or lower than in 2000. However, the chance of obtaining a difference as large as that observed is not that unusual (more than 5%) if, in fact, there is no real difference between the average scores in 2005 and 2000. In simple terms, what we are observing may just be due to chance variation, and we would not take this as strong evidence that the average score in 2005 is different from 2000.

Exercise 6.71

(a) The researchers want to determine if there is a difference between the computer's and the driver's calculations of mpg; that is, does the mean of the differences differ from 0. The alternative is $\mu \neq 0$, so the hypotheses of interest to the researchers are H_0: $\mu = 0$ and H_a: $\mu \neq 0$.

(b) The value of the test statistic is $z = \dfrac{\bar{x} - \mu_0}{\sigma / \sqrt{n}} = \dfrac{2.73 - 0}{3 / \sqrt{20}} = 4.07$. Since the alternative is $\mu \neq 0$, the P-value is the chance of getting a value of $\bar{x}$ at least 2.73 mpg from zero in either direction if the true mean were 0. In terms of the test statistic, this is equivalent to computing $2P(Z \geq |z|) = 2P(Z \geq 4.07) \approx 0$.

There is very strong evidence that the mean score of the differences is different from zero, and in this case greater than zero. The computer's calculations of mpg is systematically larger than the driver's calculations.

Exercise 6.73

(a) The advocacy group wants to determine if the mean nicotine content is higher than 1.4. The alternative is $\mu > 1.4$, so the hypotheses of interest to the advocacy group are H_0: $\mu = 0$ and H_a: $\mu > 1.4$.

(b) The value of the test statistic is given as $z = 2.36$. The P-value is $P(Z > 2.36) = 1 - 0.9909 = 0.0091$. Since the value is below 0.05, we reject at the 5% level of significance.

Using Table D, we look up the critical value corresponding to the tail probability 0.05 and find the critical value $z^* = 1.645$. A 5% level test rejects if $z > 1.645$. Since 2.36 exceeds this value, we reject at the 5% level of significance.

(b) The *P*-value was computed in (a) to be 0.0091. Since 1% = 0.01 is greater than this value, we reject at the 1% level of significance. This is the advantage of giving the *P*-value. It allows us to assess significance at any level.

Using Table D, we look up the critical value corresponding to the tail probability 0.01 and find the critical value $z^* = 2.326$. Since $z = 2.36$ is greater than this critical value, we reject at the 1% level of significance.

Exercise 6.79

The two-sided test at the 0.5% level rejects whenever $z < -2.807$ or $z > 2.807$ and the two-sided test at the 0.1% level rejects whenever $z < -3.291$ or $z > 3.291$. If you chose any value of z for which $-3.291 < z < -2.807$ or $2.807 < z < 3.291$ then the test gives a significant result at the 0.5% level but not at the 0.1% level.

SECTION 6.3

OVERVIEW

When describing the outcome of a hypothesis test it is more informative to give the *P*-value than just the reject or fail to reject decision at a particular significance level α. The traditional levels of 0.01, 0.05, and 0.10 are arbitrary and serve as rough guidelines.

When testing hypotheses with a very large sample, the *P*-value can be very small for effects that may not be of practical interest. Don't confuse small *P*-values with large or important effects. Plot the data to display the effect you are trying to show, and also give a confidence interval, which says something about the size of the effect.

Just because a test is not statistically significant doesn't imply that the null hypothesis is true. This may occur when the test is based on a small sample size and has low power. Finally, if you run enough tests, you will invariably find statistical significance for one of them. Be careful in interpreting the results when testing many hypotheses on the same data.

GUIDED SOLUTIONS

Exercise 6.95

KEY CONCEPTS: Statistical significance and sample size

For testing the hypotheses $H_0: \mu = 505$ and $H_a: \mu > 505$, the significance test is based on the z statistic, $z = \dfrac{\bar{x} - \mu_0}{\sigma / \sqrt{n}}$. In each of the three cases, $\bar{x} = 508$, $\mu_0 = 505$, and $\sigma = 100$. The only thing changing is the value of n which is 100, 1000, and 10,000. Give the test statistic for each sample size in the following table and the *P*-value. How does the *P*-value change? What does this say about the effect of sample size on declaring very small effects statistically significant?

Sample size	Test statistic	*P*-value
$n = 100$	$z = \dfrac{\overline{x} - \mu_0}{\sigma / \sqrt{n}} =$	
$n = 1000$	$z = \dfrac{\overline{x} - \mu_0}{\sigma / \sqrt{n}} =$	
$n = 10{,}000$	$z = \dfrac{\overline{x} - \mu_0}{\sigma / \sqrt{n}} =$	

Exercise 6.96

KEY CONCEPTS: Statistical significance vs. practical importance, confidence intervals

The form of the 99% confidence interval is $\overline{x} \pm z^* \dfrac{\sigma}{\sqrt{n}}$. In each of the three cases, $\overline{x} = 508$, $z^* = 2.576$, and $\sigma = 100$. The only thing changing is the value of n which is 100, 1000, and 10,000. Give the three intervals in the table below. How do the intervals change? How small can the effect of coaching be?

Sample size	Confidence interval
$n = 100$	$\overline{x} \pm z^* \dfrac{\sigma}{\sqrt{n}} =$
$n = 1000$	$\overline{x} \pm z^* \dfrac{\sigma}{\sqrt{n}} =$
$n = 10{,}000$	$\overline{x} \pm z^* \dfrac{\sigma}{\sqrt{n}} =$

Exercise 6.100

KEY CONCEPTS: Significance levels vs. *P*-values

We are going to compare the two groups: those who eventually become executives and those who don't succeed and leave the company. In addition, suppose we have measured 100 variables on these individuals and are going to compare the groups on each of these 100 variables. That is, we are going to do an appropriate statistical test at the 0.05 significance level to compare the two groups for each of the 100 variables and we run 100 tests in all. About how many tests would lead to statistical significance if none of the variables had any real effects? Does this create a problem with using the "significant" variables to select future trainees? How might you conduct a follow-up study to clarify the importance of the variables identified by the first study?

Exercise 6.105

KEY CONCEPTS: Multiple tests, the Bonferroni procedure

If you perform k tests and want protection at level α, use α/k as your cutoff for statistical significance. In our case, $k = 12$ and $\alpha = 0.05$, so each test is conducted at the $0.05/7 = 0.00417$ significance level. Which *P*-values given lead to rejection at the 0.00417 level?

COMPLETE SOLUTIONS

Exercise 6.95

The three z statistics and *P*-values are given below.

Sample size	Test statistic	*P*-value
$n = 100$	$z = \dfrac{\bar{x} - \mu_0}{\sigma / \sqrt{n}} = \dfrac{508 - 505}{100 / \sqrt{100}} = 0.3$	$P(Z \geq z) = P(Z \geq 0.3) = 0.3821$
$n = 1000$	$z = \dfrac{\bar{x} - \mu_0}{\sigma / \sqrt{n}} = \dfrac{508 - 505}{100 / \sqrt{1000}} = 0.95$	$P(Z \geq z) = P(Z \geq 0.95) = 0.1711$
$n = 10{,}000$	$z = \dfrac{\bar{x} - \mu_0}{\sigma / \sqrt{n}} = \dfrac{508 - 505}{100 / \sqrt{10000}} = 3.0$	$P(Z \geq z) = P(Z \geq 3.0) = 0.0013$

You can see that the z statistics are getting larger with increasing sample size, and the *P*-values are decreasing. The last confidence test shows a statistically significant result, namely that the mean score is higher than 505. A large sample size will declare very small effects statistically significant because the observed effect cannot easily be explained by chance, despite the fact it is small. The advantage of confidence intervals, over just reporting *P*-values or statistical significance, is that a confidence interval tells something about the size of the effect, not just whether it is statistically significant.

Exercise 6.96

The three confidence intervals are given below.

Sample size	Confidence interval
$n = 100$	$\bar{x} \pm z^* \frac{\sigma}{\sqrt{n}} = 508 \pm 2.576 \frac{100}{\sqrt{100}} = 508 \pm 25.76 = (482.24, 533.76)$
$n = 1000$	$\bar{x} \pm z^* \frac{\sigma}{\sqrt{n}} = 508 \pm 2.576 \frac{100}{\sqrt{1000}} = 508 \pm 8.15 = (499.85, 516.15)$
$n = 10{,}000$	$\bar{x} \pm z^* \frac{\sigma}{\sqrt{n}} = 508 \pm 2.576 \frac{100}{\sqrt{10000}} = 508 \pm 2.576 = (505.42, 510.58)$

In the last case we see that the mean score appears to be higher after coaching, since the interval doesn't include 505 (you need to be careful relating these intervals to the one-sided tests in Exercise 6.95 since the relationship we have studied is between a confidence interval and a *two-sided* test). You can see that the intervals are getting shorter with increasing sample size, and that the last confidence interval shows the mean score to be higher than 505, but only by a very small amount. The advantage of confidence intervals, over just reporting *P*-values or statistical significance, is that a confidence interval tells something about the size of the effect, not just whether it is statistically significant.

Exercise 6.100

If none of the variables had any real effects and we ran 100 tests, we would expect to find a statistically significant difference between those who eventually become executives and those who don't on about 5 variables, just by chance. It would be better to use the variables that appeared most promising in terms of statistical significance in a follow-up study that concentrated on the effects of just these variables when comparing groups.

Exercise 6.105

If the *P*-value is less than $\alpha/k = 0.05/12 = 0.00417$, then we reject using the Bonferroni procedure. In this case we reject the null hypothesis for three of the tests – the ones with *P*-values of 0.001 (the fifth), 0.004 (the sixth), and 0.002 (the eleventh), because these *P*-values are smaller than 0.00417.

SECTION 6.4

OVERVIEW

The **power** of a significance test is always calculated for a specific alternative hypothesis and is the probability that the test will reject H_0 when that alternative is true. This calculation requires knowledge of the sampling distribution under the specific alternative hypothesis of the test statistic used. Power is usually interpreted as the ability of a test to detect an alternative hypothesis or as the sensitivity of a test to an alternative hypothesis. The power of a test can be increased by increasing the sample size when the significance level remains fixed.

To compute the power of a significance test for a mean of a normal population, we need to

- state H_0, H_a (the particular alternative we want to detect), and the significance level α,
- find the values of $\overline{x}$ that will lead us to reject H_0,
- calculate the probability of observing these values of $\overline{x}$ when the alternative is true.

Statistical inference can be regarded as giving rules for making decisions in the presence of uncertainty. From this **decision theory** point of view, H_0 and H_a are just two statements of equal status that we must decide between. Decision analysis chooses a rule for deciding between H_0 and H_a on the basis of the probabilities of the two types of errors that we can make. A **Type I error** occurs if H_0 is rejected when it is in fact true. A **Type II error** occurs if H_0 is accepted when in fact H_a is true.

There is a clear relation between α level significance tests and testing from the decision making point of view. The significance level α is the probability of a Type I error, and the power of the test to detect a specific alternative is 1 minus the probability of a Type II error for that alternative.

GUIDED SOLUTIONS

Exercise 6.113

KEY CONCEPTS: Power of a significance test

To compute the power of a significance test about a mean, we need to

(i) state H_0, H_a (the particular alternative we want to detect), and the significance level α,

(ii) find the values of $\overline{x}$ that will lead us to reject H_0,

(iii) calculate the probability of observing these values of $\overline{x}$ when the alternative is true.

Following these steps we notice the following.

(i) In this problem the hypotheses are

$$H_0: \mu = 450 \text{ vs. } H_a: \mu > 450$$

The particular alternative we want to detect is $\mu = 460$. The significance level is $\alpha = 0.01$.

(ii) The values of $\overline{x}$ that will lead us to reject H_0 are indicated in the problem and are those for which

$$z = \frac{\overline{x} - 450}{100/\sqrt{500}} \geq 2.326$$

since $\sigma = 100$ and the sample size $n = 500$. Express the above inequality in terms of values of $\bar{x}$ by filling in the blank below.

We reject H_0 if $\bar{x} \geq$ __________.

(iii) The probability of observing these values of $\bar{x}$ when the alternative is true is

$$P(\bar{x} \geq 460.4 \text{ when } \mu = 460) = P\left(\frac{\bar{x}-\mu}{\sigma/\sqrt{n}} \geq \frac{460.4-460}{100/\sqrt{500}}\right) =$$

Now complete the calculation of this probability using Table A. The result will be the desired power.

Exercise 6.115

KEY CONCEPTS: Type I and Type II errors

(a) We suppose that the automated diagnostic procedure will clear the patient if the procedure does not indicate the presence of a specific disease and will refer the patient to a doctor if the procedure indicates the presence of a disease. What would the null and alternative hypotheses be? Write them in terms of whether or not the patient has a disease.

H_0: $\qquad\qquad\qquad\qquad$ H_a:

With these hypotheses, relate the two types of errors in terms of "false-positive" and "false-negative" test results.

(b) The answer to this question depends on the seriousness of the disease as well as the consequences of treatment or additional tests. Give your answer in terms of these.

COMPLETE SOLUTIONS

Exercise 6.113

We follow the three steps indicated in the Guided Solutions.

(i) This step was completely discussed in the Guided Solutions.
(ii) In terms of $\bar{x}$, after solving the inequality given in the Guided Solutions, we reject H_0 if $\bar{x} \geq (2.326 \times 100/\sqrt{500}) + 450 = 460.4$.

(iii) We find

$$P(\bar{x} \geq 460.4 \text{ when } \mu = 460) = P\left(\frac{\bar{x} - \mu}{\sigma / \sqrt{n}} \geq \frac{460.4 - 460}{100 / \sqrt{500}}\right)$$

$$= P(Z \geq 0.09) = 0.4641$$

where we have used Table A to compute $P(Z \geq 0.09)$. The desired power is 0.4641. The probability of detecting an increase of 10 points is only 0.4641, which may not be as large as the experimenter would like. To remedy this, a larger sample would need to be taken.

Exercise 6.115

(a) The hypotheses would be

H_0: Patient does not have a disease $\qquad$ H_a: Patient has a disease

With these hypotheses, a Type I error, which indicates rejection of the null hypothesis when it is true, corresponds to a "false-positive" test result, that is, declaring a well patient to have a disease. A Type II error corresponds to accepting the null hypothesis when it is false and corresponds to a "false-negative" test result, that is, declaring a patient with a disease to be well.

(b) If a disease were very serious, such as a terminal disease that requires fast treatment, then we would want the probability of a Type II error, or "false-negative" test result, to be very small. That is, we wouldn't like to miss the disease and have possibly fatal consequences. For more trivial health problems that might have obvious symptoms occurring later and for which treatment at that later point would still be effective, we might prefer to keep the probability of a Type I error, or "false-positive," small.

CHAPTER 7

INFERENCE FOR DISTRIBUTIONS

SECTION 7.1

OVERVIEW

Confidence intervals and significance tests for the mean μ of a normal population are based on the sample mean $\overline{x}$ of an SRS. When the sample size n is large, the central limit theorem suggests that these procedures are approximately correct for other population distributions. In Chapter 6, we considered the (unrealistic) situation in which we knew the population standard deviation σ. In this section, we consider the more realistic case where σ is not known and we must estimate σ from our SRS by the sample standard deviation s. In Chapter 6 we used the **one-sample z statistic**

$$z = \frac{x - \mu}{\sigma / \sqrt{n}}$$

which has the $N(0,1)$ distribution. Replacing σ by s, we now use the **one-sample t statistic**

$$t = \frac{\overline{x} - \mu}{s / \sqrt{n}}$$

which has the ***t* distribution** with $n - 1$ **degrees of freedom.**

For every positive value of k, there is a t distribution with k degrees of freedom, denoted $t(k)$. All are symmetric, bell-shaped distributions, similar in shape to normal distributions but with greater spread. As k increases, $t(k)$ approaches the $N(0,1)$ distribution.

A level C **confidence interval for the mean** μ of a normal population when σ is unknown is

$$\overline{x} \pm t^* \frac{s}{\sqrt{n}}$$

where t^* is the upper $(1 - C)/2$ critical value of the $t(n - 1)$ distribution, whose value can be found in Table D of the text or from statistical software. The quantity $t^* \frac{s}{\sqrt{n}}$ is the **margin of error.**

Significance tests of H_0: $\mu = \mu_0$ are based on the one-sample t statistic. Such tests are often referred to as **one-sample *t* tests.** *P*-values or fixed significance levels are computed from the $t(n-1)$ distribution using Table D or, more commonly in practice, using statistical software.

The power of the t test is calculated like that of the z test as described in Chapter 6, using an approximate value (perhaps based on past experience or a pilot study) for both σ and s.

One application of these one-sample t procedures is to the analysis of data from **matched pairs** studies. We compute the difference between the two values of a matched pair (often before-and-after measurements on the same unit) to produce a single sample value. We do this for each of the matched pairs. The sample mean and standard deviation of these differences are computed. Depending on whether we are interested in a confidence interval or a test of significance concerning the population mean of matched pairs, we use either the one-sample confidence interval or the one-sample significance test based on the t statistic.

For larger sample sizes, the t procedures are fairly **robust** against nonnormal populations. As a rule of thumb, t procedures are useful for nonnormal data when $n \geq 15$ unless the data show outliers or strong skewness, and for samples of size $n \geq 40$ t procedures can be used for even clearly skewed distributions. Data consisting of small samples from skewed populations can sometimes be analyzed by first applying a **transformation** (such as logarithms) to the data to obtain an approximately normally distributed variable. The t procedures can then be applied to the transformed data. When transformations are used, it is a good idea to examine stemplots, histograms, or normal quantile plots of the transformed data to verify that the transformed data appear to be approximately normally distributed.

Another procedure that can be used with smaller samples from a nonnormal population is the **sign test.** The sign test is most useful for testing for "no treatment effect" in matched pairs studies. As with the matched pairs t test, one computes the differences of the two values in each matched pair. Pairs with difference 0 are ignored and the number of trials n is the count of the remaining pairs. The test statistic is the count X of pairs with a positive difference. *P*-values for X are based on the binomial $B(n, 1/2)$ distribution. The sign test is less powerful than the t test in cases where the use of the t test is justified.

GUIDED SOLUTIONS

Exercise 7.25

KEY CONCEPTS: Confidence intervals based on the one-sample t statistic, checking assumptions

(a) A stemplot is most easily constructed using statistical software. Either use statistical software or complete the stemplot below. We have used split stems. Does the distribution appear skewed? Are there any obvious outliers? A boxplot is provided on the next page.

```
0 |
0 |
1 |
1 |
2 |
2 |
3 |
3 |
4 |
4 |
5 |
5 |
6 |
6 |
```

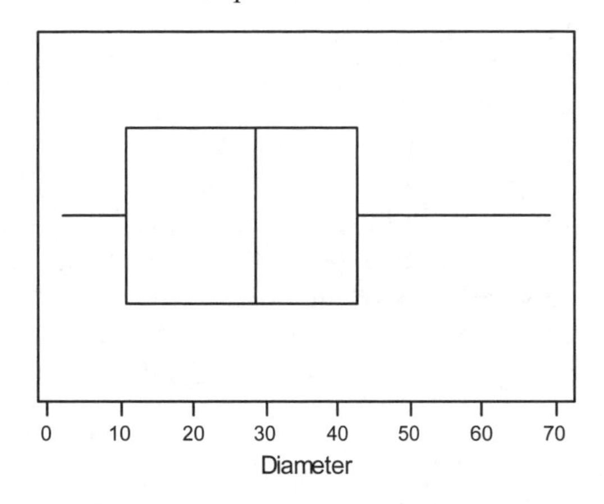

(b) Can the methods of this section be applied to find a 95% confidence interval for the mean DBH of all trees in the Wade Tract? Why or why not?

(c) To compute a 95% confidence interval we use the formula

$$\bar{x} \pm t^* \frac{s}{\sqrt{n}}$$

where t^* is the upper $(1 - C)/2$ critical value of the $t(n - 1)$ distribution. Here C is the confidence level of 0.95, and n is the sample size of 40. Thus, in this exercise, t^* is the upper $(1 - 0.95)/2 = 0.025$ critical value of the $t(40 - 1) = t(39)$ distribution. Approximate this critical value in Table D or use software to find the critical value.

$t^* =$

Next compute $\bar{x}$ and the margin of error $t^* \frac{s}{\sqrt{n}}$, preferably using statistical software or a calculator.

$\bar{x} =$ $t^* \frac{s}{\sqrt{n}} =$

Finally, substitute all these values into the formula to complete the computation of the 95% confidence interval.

$\bar{x} \pm t^* \frac{s}{\sqrt{n}} =$

Write a summary describing the meaning of the confidence interval.

Note that a much easier approach would be to use statistical software to compute the 95% confidence interval directly using the raw data. Consult the manual for your software for instructions.

(d) Would these results apply to other similar trees in the same area? Explain.

Exercise 7.35

KEY CONCEPTS: Matched pairs experiments, one-sample t tests

(a) This is a matched pairs experiment. Each time the gas tank is filled, a pair of measurements is taken. The first is a computer calculation of the miles per gallon between fill-ups and the second is a driver calculation based on the number of miles driven divided by the number of gallons added to the tank between fill-ups. The driver is interested in determining if these calculations are different. Letting μ correspond to the mean difference in calculations, set up the null and alternative hypotheses.

H_0:
H_a:

(b) The first step in the analysis of a matched pairs experiment is to take the differences between the two measurements within each pair. You can do this on a computer or fill in the differences below. To be consistent with the complete solution, take the differences as

(computer measurement) – (driver measurement)

Fill-up	1	2	3	4	5	6	7	8	9	10
Computer	41.5	50.7	36.6	37.3	34.2	45.0	48.0	43.2	47.7	42.2
Driver	36.5	44.2	37.2	35.6	30.5	40.5	40.0	41.0	42.8	39.2
Difference										

Fill-up	11	12	13	14	15	16	17	18	19	20
Computer	43.2	44.6	48.4	46.4	46.8	39.2	37.3	43.5	44.3	43.3
Driver	38.8	44.5	45.4	45.3	45.7	34.2	35.2	39.8	44.9	47.5
Difference										

Use a computer or calculator to compute the mean and standard deviation of the differences and fill in your answers below.

$\overline{x}$ = s =

Use this information to calculate the one-sample t statistic:

$$t = \frac{\overline{x} - \mu}{s/\sqrt{n}} =$$

Based on the value of this t statistic, between which levels from Table D does the P-value lie? What conclusions do you draw?

Exercise 7.36

KEY CONCEPTS: One-sample t confidence intervals

(a) The six measurements on phosphate level are reproduced below in mg/dl.

5.6 5.1 4.6 4.8 5.7 6.4

Use a computer or calculator to compute the mean, standard deviation, and standard error of the mean for these 6 phosphate measurements, and fill in your answers below.

$\overline{x} =$ $s =$ $\frac{s}{\sqrt{n}} =$

(b) Use your answers in (a) to give a 90% confidence interval for this patient's mean phosphate level.

$$\overline{x} \pm t^* \frac{s}{\sqrt{n}} =$$

Exercise 7.37

KEY CONCEPTS: One-sample t tests

We are interested in whether or not the patient's mean level μ falls above 4.8. What effect do you wish to show? Use this to set up the null and alternative hypotheses

H_0:
H_a:

Make sure you understand why we use a one-sided alternative here.

The sample mean and standard deviation for the data given in Exercise 7.36 are easily calculated to be

$\overline{x} = 5.367$ $s = 0.6653$

and the standard error is

$$\frac{s}{\sqrt{n}} = \frac{0.6653}{\sqrt{6}} = 0.272$$

Use this information to calculate the one-sample t statistic:

$$t = \frac{\overline{x} - \mu}{s/\sqrt{n}} =$$

Based on the value of this t statistic, between which levels from Table D does the P-value lie? What conclusions do you draw?

Exercise 7.49

KEY CONCEPTS: Sign test

In Exercise 7.35 we investigated whether the mean μ for the population of differences

(computer calculation) – (driver calculation)

for matched pairs was different than 0. In this problem we are interested in a one-sided alternative. Specifically, we are interested in whether the computer calculates a higher mpg than the driver.

Try to formulate the hypotheses in terms of

- the median for the population of these differences

and in terms of

- the probability p that the computer calculates a higher mpg than the driver

What do the statements H_0: $\mu = 0$ and H_a: $\mu \neq 0$ imply about the median for the population of differences? Remember, the median and mean both attempt to measure the center of a population distribution.

Hypotheses in terms of the median: H_0:
H_a:

What do the statements H_0: $\mu = 0$ and H_a: $\mu \neq 0$ imply about the probability p of the computer calculation being higher than the driver calculation?

Hypotheses in terms of p: H_0:
H_a:

Recall that for the sign test, pairs with difference 0 are ignored and the number of trials n is the count of the remaining pairs. The test statistic is the count X of pairs with a positive difference, where the differences (changes) were computed in Exercise 7.35. P-values for X are based on the binomial $B(n, 1/2)$ distribution.

How many pairs with difference 0 are there in the data? What, therefore, is n?

$n =$

What is the observed number of pairs with a positive difference in our data?

$X =$

In terms of the probability p of the computer calculation being higher than the driver calculation, we wish to test the hypotheses H_0: $p = 1/2$ and H_a: $p > 1/2$. This implies that the P-value is

$$P(X \geq \text{observed number of pairs with a positive difference in our data})$$

Compute this probability using Table C. Since the sample size and success probability needed for the calculations are included in Table C there is no need to use the normal approximation to the binomial distribution to compute the P-value.

COMPLETE SOLUTIONS

Exercise 7.25

(a) A stemplot of the data is given below. DBH is fairly uniformly distributed between 2 and 52 meters, with only one potential outlier, although it is not even considered a mild outlier according to the boxplot and the outlier criteria given in Chapter 1.

```
0 | 222244
0 | 579
1 | 0113
1 | 678
2 | 2
2 | 6679
3 | 112
3 | 5789
4 | 0033444
4 | 7
5 | 112
5 |
6 |
6 | 9
```

(b) The sample of 40 trees is a random sample from the population of 584 trees in the tract. The t procedures can be used safely for a sample size of 40, even for skewed distributions. In this case there is little evidence of skewness, so the procedures can be used safely. However, we are sampling a fairly large percentage of the population, about 10%, but this should have a small effect.

(c) The desired quantities are

$$t^* = 2.0227 \qquad \bar{x} = 27.29 \qquad s = 17.71 \qquad \text{margin of error} = t^* \frac{s}{\sqrt{n}} = (2.0227)\frac{17.71}{\sqrt{40}} = 5.66$$

and $$\bar{x} \pm t^* \frac{s}{\sqrt{n}} = 27.29 \pm 5.66 = (21.63, 32.95)$$

We are 95% confident that the mean DBH of all 584 trees in the Wade Tract is between 21.63 and 32.95 meters. Although the sample size is fairly large, the interval is still quite wide because of the large variability in DBH from tree to tree in the Wade Tract.

(d) The Wade Tract is an old-growth forest of Longleaf pine trees that has survived in a relatively undisturbed state since before the settlement of the area by Europeans. One would need to be careful extrapolating these results to other similar trees in the same area, since the Wade Tract is a fairly special forest that could potentially have a different distribution of DBH values than other seemingly similar trees in the same area.

Exercise 7.35

(a) Letting μ correspond to the mean difference in calculations, the hypotheses are

$$H_0: \mu = 0$$
$$H_a: \mu \neq 0$$

since we are interested in a difference in either direction.

(b) The original measurements and the differences are reproduced below.

Fill-up	1	2	3	4	5	6	7	8	9	10
Computer	41.5	50.7	36.6	37.3	34.2	45.0	48.0	43.2	47.7	42.2
Driver	36.5	44.2	37.2	35.6	30.5	40.5	40.0	41.0	42.8	39.2
Difference	5.0	6.5	-0.6	1.7	3.7	4.5	8.0	2.2	4.9	3.0

Fill-up	11	12	13	14	15	16	17	18	19	20
Computer	43.2	44.6	48.4	46.4	46.8	39.2	37.3	43.5	44.3	43.3
Driver	38.8	44.5	45.4	45.3	45.7	34.2	35.2	39.8	44.9	47.5
Difference	4.4	0.1	3.0	1.1	1.1	5.0	2.1	3.7	-0.6	-4.2

Using the differences, the mean and standard deviation of the differences are

$\bar{x} = 2.730$ $\quad\quad$ $s = 2.802$

and the numerical value of the one-sample t statistic is

$$t = \frac{\bar{x} - \mu}{s/\sqrt{n}} = \frac{2.730 - 0}{2.802/\sqrt{20}} = 4.36$$

Using Table D with $n - 1 = 19$ degrees of freedom, we see the P-value is less than 0.0005 since the t statistic value of 4.36 is larger than the largest t^* value in the table. There is very strong evidence of a difference in the mean mpg calculation by the two methods. In this case, we see that the computer calculation tends to give larger values. A confidence interval would give further information on the size of the mean difference.

Exercise 7.36

(a) The mean, standard deviation, and standard error of the mean for the six measurements on the level of phosphate are

$\bar{x} = 5.367$ $\quad\quad$ $s = 0.6653$ $\quad\quad$ $\frac{s}{\sqrt{n}} = \frac{0.6653}{\sqrt{6}} = 0.272$

(b) The 90% confidence interval for this patient's mean phosphate level

$$\bar{x} \pm t^* \frac{s}{\sqrt{n}} = 5.367 \pm (2.015)(0.272) = 5.367 \pm 0.548 = (4.819, 5.914)$$

where $t^* = 2.015$ is found in Table D using $n - 1 = 5$ degrees of freedom.

Exercise 7.37

The hypotheses we wish to test are

$$H_0: \mu = 4.8$$
$$H_a: \mu > 4.8$$

We compute

$$t = \frac{\bar{x} - \mu}{s / \sqrt{n}} = \frac{5.367 - 4.8}{0.272} = 2.085$$

According to the row corresponding to $n - 1 = 5$ in Table D, the P-value lies between 0.025 and 0.05. Using statistical software, we find the P-value is 0.0457. We conclude that there is good evidence that the patient's mean level μ in fact falls above 4.8.

Exercise 7.49

In terms of the median for the population of differences

(computer measurement) – (driver measurement)

the hypotheses are

H_0: population median = 0 and H_a: population median > 0

since these hypotheses test whether the computer calculates a higher mpg than the driver.

In terms of the probability p of the computer calculation being higher than the driver calculation, the hypotheses would be

$$H_0: p = 1/2 \text{ and } H_a: p > 1/2$$

There are 0 pairs with difference 0 in the data; thus, $n = 20$. The number of pairs with a positive difference in our data is $X = 17$. Since X has the binomial $B(20, 1/2)$ distribution U, using Table C the P-value is

$$P\text{-value} = P(X \geq 17) = 0.0011 + 0.0002 = 0.0013$$

There is very strong evidence that the computer calculates a higher mpg than the driver.

SECTION 7.2

OVERVIEW

One of the most commonly used significance tests is the **comparison of two population means,** μ_1 and μ_2. In this setting we have two distinct, independent SRSs from two populations or two treatments on two samples. The procedures are based on the difference $\bar{x}_1 - \bar{x}_2$. When the populations are not normal, the results obtained using the methods of this section are approximately correct due to the central limit theorem.

Tests and confidence intervals for the difference in the population means, $\mu_1 - \mu_2$, are based on the **two-sample *t* statistic:**

$$t = \frac{(\bar{x}_1 - \bar{x}_2) - (\mu_1 - \mu_2)}{\sqrt{\frac{s_1^2}{n_1} + \frac{s_2^2}{n_2}}}$$

Despite the name, this test statistic does *not* have an exact *t* distribution. However, there are good approximations to its distribution that allow us to carry out valid significance tests. Conservative procedures use the $t(k)$ distribution as an approximation, where the degrees of freedom k is taken to be the smaller of $n_1 - 1$ and $n_2 - 1$. More accurate procedures use the data to estimate the degrees of freedom k; these are the methods followed by most statistical software.

To carry out a significance test of $H_0 : \mu_1 = \mu_2$, use the two-sample *t* statistic

$$t = \frac{\bar{x}_1 - \bar{x}_2}{\sqrt{\frac{s_1^2}{n_1} + \frac{s_2^2}{n_2}}}$$

The *P*-value is found by using the approximate distribution $t(k)$, where k is estimated from the data when using statistical software or is the smaller of $n_1 - 1$ and $n_2 - 1$ for a conservative procedure.

An approximate confidence level *C* **confidence interval** for $\mu_1 - \mu_2$ is given by

$$(\bar{x}_1 - \bar{x}_2) \pm t^* \sqrt{\frac{s_1^2}{n_1} + \frac{s_2^2}{n_2}}$$

where t^* is the upper $(1 - C)/2$ critical value for $t(k)$, where k is estimated from the data when using statistical software or is the smaller of $n_1 - 1$ and $n_2 - 1$ for a conservative procedure. The procedures are most robust to failures in the assumptions when the sample sizes are equal.

The **pooled two-sample *t* procedures** are used when we can safely assume that the two populations have equal variances. The modifications in the procedures are the use of the pooled estimator of the common unknown variance

$$s_p^2 = \frac{(n_1 - 1)s_1^2 + (n_2 - 1)s_2^2}{n_1 + n_2 - 2}$$

and critical values obtained from the $t(n_1 + n_2 - 2)$ distribution.

GUIDED SOLUTIONS

Exercise 7.81

KEY CONCEPTS: Tests and confidence intervals using the two-sample *t*, side-by-side boxplots, summary statistics

(a) One of the simplest ways to compare two distributions is to draw side-by-side boxplots. This was discussed in Section 1 of Chapter 1 and you may want to review that material to help you interpret the boxplots. Side-by-side boxplots give a quick way of determining if the spreads of the two groups are similar, how the locations compare, and if there are any outliers in either group. With back-to-back stemplots, one needs to be a little careful when comparing data displays for two groups with unequal sample sizes because the different sample sizes could affect your visual impression of the two plots. In this exercise, the sample sizes are equal, so this would not be an issue. We have used software to draw the side-by-side boxplots and have displayed them below.

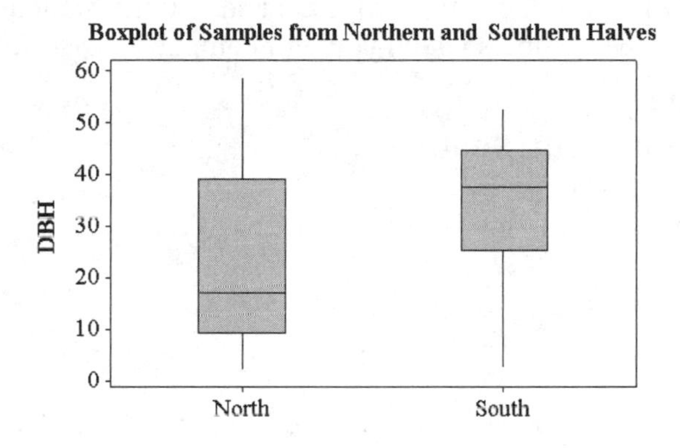

Complete the back-to-back stemplot below. We have rounded all measurements to the nearest integer with the stems being the 10's and the leaves being the 1's. Split stems are used to get a clearer plot. We have included the first three rounded measurements from the North (28, 15, and 39) and the first three rounded measurements from the South (44, 26, and 50) to get you started.

South		North
	0	
	0	
	1	
	1	5
	2	
6	2	8
	3	
	3	9
4	4	
	4	
0	5	
	5	

Describe the patterns in the data that you observe in the two plots.

(b) Is it appropriate to use the two-sample t procedures to compare the mean DBH of the trees in the northern and southern halves? Review the guidelines on page 456.

(c) Let μ_N denote the mean DBH for the Northern trees and μ_S denote the mean DBH for the Southern trees. State the hypotheses of interest in terms of these parameters.

H_0 : H_a :

(d) When means and standard deviations for the two samples are given, it is relatively easy to compute the value of the two-sample t statistic. With raw data as in this exercise, the details of the computations are best left to statistical software. However, although you may let the computer do the tedious computations, it is still important to know what all the numbers in the output mean and how to interpret the results. The output below is reproduced from Minitab and is fairly standard among different software packages. We have omitted some numbers which you will compute yourself. The full output is provided in the Complete Solutions.

Two-Sample T-Test and CI: North, South

```
Two-sample T for North vs South

        N  Mean  StDev  SE Mean
North  30  23.7   17.5      3.2
South  30  34.5   14.3      2.6


Difference = mu (North) - mu (South)
Estimate for difference:  -10.83
95% CI for difference:
T-Test of difference = 0 (vs not =): T-Value = ______  P-Value = ______  DF = ___
```

The output begins with summary information for the two samples. For the northern trees, $n_N = 30$, $\bar{x}_N = 23.7$, $s_N = 17.5$, and SE Mean is $s_N / \sqrt{n_N} = 3.2$. For the southern trees, $n_S = 30$, $\bar{x}_S = 34.5$, $s_S = 14.3$, and SE Mean is $s_S / \sqrt{n_S} = 2.6$.

The last line of the output gives the hypotheses, test statistic, P-value, and approximate degrees of freedom to the nearest integer below the computed value. The "`T-Test of difference = 0 (vs not =)`" indicates the direction of the alternative is two-sided. Some software packages allow you to specify a one-sided alternative while others automatically test a two-sided alternative. Be careful when using the reported P-value if your software automatically computes a two-sided test but you are interested in a one-sided test.

Now, compute the T-value which has been omitted using the summary information in the beginning of the output.

$$t = \frac{\bar{x}_1 - \bar{x}_2}{\sqrt{\frac{s_1^2}{n_1} + \frac{s_2^2}{n_2}}} =$$

The degrees of freedom is found using the formula

$$\frac{\left(\frac{s_N^2}{n_N}+\frac{s_S^2}{n_S}\right)^2}{\frac{1}{n_N-1}\left(\frac{s_N^2}{n_N}\right)^2+\frac{1}{n_S-1}\left(\frac{s_S^2}{n_S}\right)^2}=$$

Use your value of the t statistic and the degrees of freedom to approximate the P-value. Is the mean DBH of trees different for the northern half than the southern half at the 5% level? At the 1% level? Why? Summarize your conclusions.

(e) Using the summary information in the output, find a 95% confidence interval for the difference in mean DBHs. The formula for the confidence interval is

$$(\overline{x}_N-\overline{x}_S)\pm t^*\sqrt{\frac{s_N^2}{n_N}+\frac{s_S^2}{n_S}}=$$

where $t^* = 2.003$ is the $(1 - C)/2 = 0.025$ critical value for the t distribution with degrees of freedom equal to 55.8. You can obtain this critical value from statistical software or approximate it using Table D.

Exercise 7.85

KEY CONCEPTS: Tests and confidence intervals using the two-sample t, conservative degrees of freedom

(a) Let μ_B and μ_F represent the population mean hemoglobin levels for breast-fed and formula-fed infants, respectively. First state the hypotheses of interest in terms of these parameters.

$H_0:$ $\qquad$ $H_a:$

The summary statistics are reproduced below. Use these summary statistics to compute the value of t.

Group	n	$\overline{x}$	s
Breast-fed	23	13.3	1.7
Formula	19	12.4	1.8

$$t=\frac{\overline{x}_B-\overline{x}_F}{\sqrt{\frac{s_B^2}{n_B}+\frac{s_F^2}{n_F}}}=$$

Using the conservative degrees of freedom, what can you say about the P-value? What do you conclude about the hemoglobin levels of breast-fed vs. formula-fed infants?

(b) The formula for the 95% confidence interval is $(\bar{x}_1 - \bar{x}_2) \pm t^* \sqrt{\frac{s_1^2}{n_1} + \frac{s_2^2}{n_2}}$, where t^* is the upper $(1 - C)/2 = 0.025$ critical value for the t distribution with degrees of freedom equal to the smaller of $n_W - 1$ and $n_M - 1$. The means and standard deviations are given in part (a). Don't forget to square the standard deviations in the formula for the standard error. Complete the calculations in steps as suggested below.

$t^* =$

$\sqrt{\frac{s_B^2}{n_B} + \frac{s_F^2}{n_F}} =$

$(\bar{x}_B - \bar{x}_F) \pm t^* \sqrt{\frac{s_B^2}{n_B} + \frac{s_F^2}{n_F}} =$

(c) State the assumptions necessary for the procedures in (a) and (b) to be valid.

COMPLETE SOLUTIONS

Exercise 7.81

(a) The data are displayed using side-by-side boxplots in the Guided Solutions and back-to back stemplots given below. Both give a similar impression of the pattern in the data. The North distribution is slightly more variable than the South distribution and the North distribution is right-skewed while the South distribution is left-skewed. There do not appear to be any outliers. Both plots suggest that the DBH scores are lower for Northern half than for the Southern half.

South		North
2	0	22334
75	0	56
2	1	01334
8	1	559
31	2	
986	2	5578
2	3	0
987665	3	699
444300	4	34
875	4	6
2110	5	4
	5	58

(b) The sample sizes are equal with the sum of the sample sizes equal to 60. Although the distributions are skewed there are no outliers and the sample sizes should be large enough for the procedures to be accurate.

(c) We are interested in whether the mean DBH scores are different for the Northern and Southern halves. This is a two-sided alternative and the hypotheses are $H_0 : \mu_N = \mu_S$ versus $H_a : \mu_N \neq \mu_S$.

(d) The full Minitab output is produced below.

Two-Sample T-Test and CI: North, South

```
Two-sample T for North vs South

        N   Mean   StDev   SE Mean
North  30   23.7    17.5       3.2
South  30   34.5    14.3       2.6


Difference = mu (North) - mu (South)
Estimate for difference:   -10.83
95% CI for difference:   (-19.09, -2.57)
T-Test of difference = 0 (vs not =): T-Value = -2.63  P-Value = 0.011  DF = 55
```

The T-value is computed as

$$t = \frac{\bar{x}_N - \bar{x}_S}{\sqrt{\frac{s_N^2}{n_N} + \frac{s_S^2}{n_S}}} = \frac{23.7 - 34.5}{\sqrt{\frac{(17.5)^2}{30} + \frac{(14.3)^2}{30}}} = -2.62$$

which agrees with the output up to roundoff error. Although the computer output has rounded the means and standard deviations to the nearest tenth for the printout, the calculations that are used by Minitab to obtain the T-value use many more digits for the means and standard deviations. The degrees of freedom are

$$\frac{\left(\frac{s_N^2}{n_N} + \frac{s_S^2}{n_S}\right)^2}{\frac{1}{n_N - 1}\left(\frac{s_N^2}{n_N}\right)^2 + \frac{1}{n_S - 1}\left(\frac{s_S^2}{n_S}\right)^2} = \frac{\left(\frac{(17.5)^2}{30} + \frac{(14.3)^2}{30}\right)^2}{\frac{1}{30-1}\left(\frac{(17.5)^2}{30}\right)^2 + \frac{1}{30-1}\left(\frac{(14.3)^2}{30}\right)^2} = 55.8$$

and the *P*-value is reported as 0.011, so there is very strong evidence that there is a difference in mean DBH between the Northern and Southern halves. We would reject the null hypothesis at the 5% but not the 1% levels.

(e) The confidence interval is

$$(\bar{x}_N - \bar{x}_S) \pm t^* \sqrt{\frac{s_N^2}{n_N} + \frac{s_S^2}{n_S}} = (23.7 - 34.5) \pm 2.003\sqrt{\frac{(17.5)^2}{30} + \frac{(14.3)^2}{30}} = -10.8 \pm 8.3 = (-19.1, -2.5)$$

which agrees with the Minitab output. Although our result is clearly statistically significant, the confidence interval provides additional information about the size of the difference between the groups.

Exercise 7.85

(a) We test $H_0: \mu_B = \mu_F$ versus $H_a: \mu_B > \mu_F$ since we are interested in determining if there is evidence of higher hemoglobin levels among the breast-fed infants. From the summary statistics the value of t is computed as

$$t = \frac{\bar{x}_B - \bar{x}_F}{\sqrt{\frac{s_B^2}{n_B} + \frac{s_F^2}{n_F}}} = \frac{13.3 - 12.4}{\sqrt{\frac{(1.7)^2}{23} + \frac{(1.8)^2}{19}}} = 1.65$$

Since the degrees of freedom are the smaller of $n_B - 1 = 22$ and $n_F - 1 = 18$, we go to Table D using df = 18 and find the two values that bracket the computed value of $t = 1.65$.

df = 18

p	0.05	0.10
t^*	1.734	1.330

Because the test is one-sided, $0.05 < P\text{-value} < 0.10$ (computer software gives the exact P-value for t = 1.65 and 18 degrees of freedom as 0.058). This gives some evidence that breast-fed infants have higher hemoglobin levels than formula-fed infants.

(b) The upper $(1 - C)/2 = 0.025$ critical value for the t distribution with degrees of freedom equal to the smaller of $n_B - 1 = 22$ and $n_F - 1 = 18$ is $t^* = 2.101$. The standard error is

$$\sqrt{\frac{s_B^2}{n_B} + \frac{s_F^2}{n_F}} = \sqrt{\frac{(1.7)^2}{23} + \frac{(1.8)^2}{19}} = 0.544$$

and the 95% confidence interval is

$$(\bar{x}_B - \bar{x}_F) \pm t^* \sqrt{\frac{s_B^2}{n_B} + \frac{s_F^2}{n_F}} = (13.3 - 12.4) \pm (2.101)(0.544) = (-0.24, 2.04)$$

(c) The total sample size is 23 + 19 = 42. The t procedures can be used even with skewed distributions with some outliers for these sample sizes. The important assumption is that the infants in the study can be considered random samples from the population of breast-fed and the population of formula-fed infants.

SECTION 7.3

OVERVIEW

There are formal inference procedures to compare the standard deviations of two normal populations as well as the two means. The validity of the procedures is seriously affected by nonnormality, and they are not recommended for regular use. The procedures are based on the ***F* statistic,** which is the ratio of the two sample variances:

$$F = \frac{s_1^2}{s_2^2}$$

If the data consist of independent simple random samples of sizes n_1 and n_2 from two normal populations, then the F-statistic has the F distribution, $F(n_1 - 1, n_2 - 1)$, if the two population standard deviations σ_1 and σ_2 are equal.

The power of the pooled two-sample t test can be found using the noncentral t distribution or a normal approximation to it. The critical value for the significance test, the degrees of freedom, and the noncentrality parameter for alternatives of interest are all required for the computation. Power calculations can be useful when comparing alternative designs and assumptions.

GUIDED SOLUTIONS

Exercise 7.105

KEY CONCEPTS: F test for equality of the standard deviations of two normal populations

The data are from Exercise 7.81 and concern the diameters of trees in the Wade Tract for random samples selected from the North and South portions of the tract. We have reproduced the summary statistics from Exercise 7.81 in the table below.

Group	n	$\bar{x}$	s
North	30	23.7	17.5
South	30	34.5	14.3

We are interested in testing the hypotheses $H_0 : \sigma_1 = \sigma_2$ and $H_a : \sigma_1 \neq \sigma_2$. The two-sided test statistic is the larger variance divided by the smaller variance. To perform the test, first find the two sample variances and take the ratio of the larger to the smaller.

$$F = \frac{\text{larger } s^2}{\text{smaller } s^2} =$$

If σ_1 and σ_2 are equal, this ratio has the $F(n_1 - 1, n_2 - 1) = F(30 - 1, 30 - 1) = F(29, 29)$ distribution. You can find F^* using software or, if you are using Table E, use the $F(30, 29)$ distribution as an approximation. What value would the ratio of variances have to exceed to reject the null hypothesis at the 5% level of significance? (Remember, it is a two-sided alternative.)

What can you say about the P-value using the information in Table E or software?

Do you have any concerns about using the F test for equality of the standard deviations for these data? Do you think the normality assumption will be satisfied? Examine the graphical displays of the data in Exercise 7.81.

Complete Solutions

Exercise 7.105

For the data collected on the Northern trees the sample standard deviation is 17.5 and the sample variance is $(17.5)^2 = 306.25$. For the data collected on the Southern trees the sample standard deviation is 14.3, and the sample variance is $(14.3)^2 = 204.49$. Computing the test statistic gives

$$F = \frac{\text{larger } s^2}{\text{smaller } s^2} = \frac{306.25}{204.49} = 1.50$$

If σ_1 and σ_2 are equal, this ratio has the $F(n_1 - 1, n_2 - 1) = F(30 - 1, 30 - 1) = F(29, 29)$ distribution. From Table E, using $F(30, 29)$ we find that the computed value of $F = 1.50$ is smaller than the smallest tabled value of $F^* = 1.62$ which corresponds to a probability of 0.10. Since it is a two-sided test we have $2 \times 0.10 < P$-value, or $0.20 < P$-value. The P-value using software is P-value $= 2 \times P(F > 1.5) = 2 \times 0.14 = 0.28$, where we have used the $F(29, 29)$ distribution.

Because both distributions show some skewness, the normality assumption is clearly not satisfied. Although the test of equality of means would be robust against a failure in the assumption of normality for these data, the test for equality of the standard deviations may not be appropriate.

CHAPTER 8

INFERENCE FOR PROPORTIONS

SECTION 8.1

OVERVIEW

In this section we consider inference about a population proportion p from an SRS of size n based on the **sample proportion** $\hat{p} = X/n$, where X is the number of "successes" (occurrences of the event of interest) in the sample. If the population is at least 10 times as large as the sample, the individual observations will be approximately independent and X will have a distribution that is approximately binomial $B(n, p)$. While it is possible to develop procedures for inference about p based on the binomial $B(n, p)$ distribution, these can be awkward to work with because of the discrete nature of the binomial distribution. When n is large, we can treat $\hat{p}$ as having a distribution that is approximately normal with mean p and standard deviation $\sqrt{p(1-p)}$.

An **approximate level C confidence interval** for p is

$$\hat{p} \pm z^{*}\mathrm{SE}_{\hat{p}}$$

where z^{*} is the upper $(1 - C)/2$ critical value of the standard normal distribution,

$$\mathrm{SE}_{\hat{p}} = \sqrt{\frac{\hat{p}(1-\hat{p})}{n}}$$

is the **standard error** of $\hat{p}$, and $m = z^{*}\mathrm{SE}_{\hat{p}}$ is the **margin of error.** Use this interval for 90%, 95%, or 99% confidence when both the number of successes and the number of failures are at least 15.

The **plus four estimate** $\tilde{p} = (X + 2)/(n + 4)$ is the sample proportion with 2 successes and 2 failures added to the data. The margin of error for confidence level C is $m = z^{*}\mathrm{SE}_{\tilde{p}}$ and the standard error of $\tilde{p}$ is given by

$$\mathrm{SE}_{\tilde{p}} = \sqrt{\frac{\tilde{p}(1-\tilde{p})}{n+4}}$$

The **level C plus four confidence interval** is

$$\tilde{p} \pm m$$

Use this interval for 90%, 95%, or 99% confidence when the sample size is at least $n = 10$.

Tests of the hypothesis H_0: $p = p_0$ are based on the **z statistic**

$$z = \frac{\hat{p} - p_0}{\sqrt{\frac{p_0(1-p_0)}{n}}}$$

with P-values calculated from the $N(0, 1)$ distribution.

The **sample size** n required to obtain a confidence interval of approximate margin of error m for a proportion is

$$n = \left(\frac{z^*}{m}\right)^2 p^*(1-p^*)$$

where p^* is a guessed value for the population proportion and z^* is the upper $(1 - C)/2$ critical value of the standard normal distribution. To guarantee that the margin of error of the confidence interval is less than or equal to m no matter what the value of the population proportion may be, use a guessed value of $p^* = 1/2$, which yields

$$n = \left(\frac{z^*}{2m}\right)^2$$

GUIDED SOLUTIONS

Exercise 8.21

KEY CONCEPTS: When to use the normal approximation to the binomial test

Recall that the rule of thumb is that the normal approximation to the binomial is appropriate if *both* $np_0 \geq 10$ and $n(1 - p_0) \geq 10$. These are the conditions that we must check in each of (a) through (d).

(a)

(b)

(c)

(d)

Exercise 8.22

KEY CONCEPTS: Testing hypotheses about a proportion

(a) In this matched pairs experiment, we let p be the probability that a randomly chosen subject prefers fresh-brewed coffee to instant coffee. We want to test the claim that the majority of people prefer the taste of fresh-brewed coffee. If the majority of people prefer the taste of fresh-brewed coffee, what does this tell you about p? Now state the hypotheses of interest in terms of p.

H_0:
H_a:

Recall that tests of the hypothesis H_0: $p = p_0$ are based on the z statistic

$$z = \frac{\hat{p} - p_0}{\sqrt{\frac{p_0(1-p_0)}{n}}}$$

Identify n, p_0, and $\hat{p}$ in this example and calculate z.

$$z = \frac{\hat{p} - p_0}{\sqrt{\frac{p_0(1-p_0)}{n}}} =$$

Once you have computed z, use it to calculate the P-value.

(b) A normal curve is sketched below. Mark the location of the z statistic computed in (a) and shade the area corresponding to the P-value.

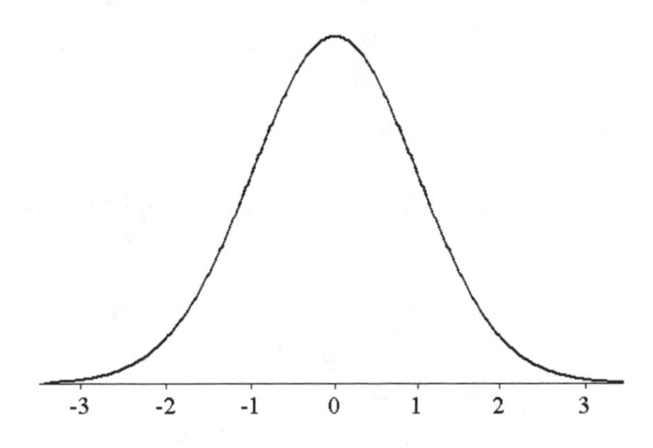

(c) Is your result significant at the 5% level? What is your practical conclusion?

Exercise 8.25

KEY CONCEPTS: Confidence intervals for a proportion

We have a 95% confidence interval, the number of successes is 594, and the number of failures is 1939. The conditions to use the large-sample confidence interval based on $\hat{p}$ are satisfied. Recall that an approximate level C confidence interval for p, the true proportion of older adults in this population who are pet owners, is

$$\hat{p} \pm z^* \sqrt{\frac{\hat{p}(1-\hat{p})}{n}}$$

where z^* is the upper $(1 - C)/2$ critical value of the standard normal distribution. From the information given in the problem, provide the values requested below.

n = sample size =

$\hat{p}$ = the proportion of older adults who are pet owners =

C = level of confidence requested =

z^* = upper $(1 - C)/2$ critical value of the standard normal distribution =

You will need to use Table D to find z^*. Now substitute these values into the formula for the confidence interval to complete the problem.

$$\hat{p} \pm z^* \sqrt{\frac{\hat{p}(1-\hat{p})}{n}} =$$

Exercise 8.31

KEY CONCEPTS: Sample size and margin of error

The sample size n required to obtain a confidence interval of approximate margin of error m for a proportion is

$$n = \left(\frac{z^*}{m}\right)^2 p^*(1-p^*)$$

where p^* is a guessed value for the population proportion and z^* is the critical value of the standard normal distribution for the desired level of confidence. Since there is no guessed value given for the proportion, in order to guarantee that the margin of error of the confidence interval is less than or equal to m no matter what the value of the population proportion may be, use a guessed value of $p^* = 0.5$. This yields

$$n = \left(\frac{z^*}{2m}\right)^2$$

To apply this formula here we must determine

m = desired margin of error =

C = desired level of confidence =

z^* = the upper $(1 - C)/2$ critical value of the standard normal distribution =

From the statement of the exercise, what are these values? Once you have determined them, use the formula to compute the required sample size n.

$$n = \left(\frac{z^*}{2m}\right)^2 =$$

Complete Solutions

Exercise 8.21

(a) We see that $np_0 = (30)(0.2) = 6 < 10$, so the normal approximation to the binomial should *not* be used in this case.

(b) We see that $np_0 = (30)(0.6) = 18 \geq 10$ and $n(1-p_0) = (30)(1 - 0.6) = (30)(0.4) = 12 \geq 10$. The normal approximation to the binomial can be used in this case.

(c) We see that $np_0 = (100)(0.5) = 50 \geq 10$ and $n(1-p_0) = (100)(1 - 0.5) = (100)(0.5) = 50 \geq 10$. The normal approximation to the binomial can be used in this case.

(d) We see that $np_0 = (200)(0.01) = 2 < 10$, so the normal approximation to the binomial should *not* be used in this case.

Exercise 8.22

(a) If the majority of people prefer the taste of fresh-brewed coffee, then $p > 0.5$. The hypotheses of interest are H_0: $p = 0.5$ and H_a: $p > 0.5$. The required quantities to compute z are

n = sample size = 40

$\hat{p}$ = the proportion who prefer fresh-brewed coffee = 28/40 = 0.7

p_0 = 0.5

giving

$$z = \frac{\hat{p} - p_0}{\sqrt{\frac{p_0(1-p_0)}{n}}} = \frac{0.7 - 0.5}{\sqrt{\frac{0.5(1-0.5)}{40}}} = 2.53$$

The *P*-value is the probability of getting a *z*-value at least as large as 2.53 and is computed as

$$P(Z > 2.53) = 1 - P(Z \leq 2.53) = 1 - 0.9943 = 0.0057$$

(b) The P-value is the area to the right of 2.53 and is shaded in the normal curve below.

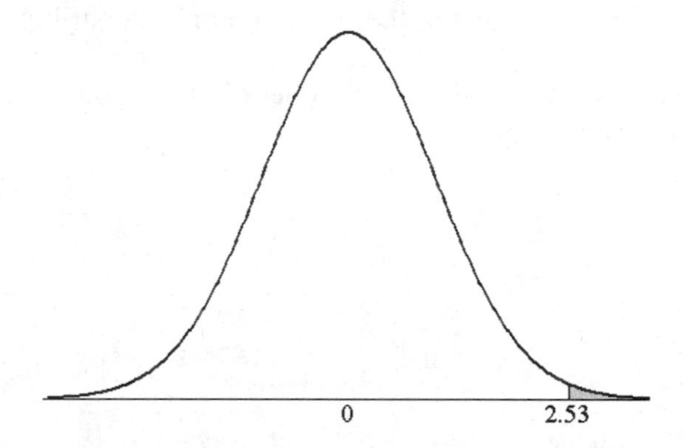

(c) We would reject at both the 5% level and the 1% level since the *P*-value is less than 0.05 and 0.01. There is strong evidence based on this experiment that the majority of people prefer freshly-brewed coffee to instant coffee. A confidence interval would give more specific information about how large a proportion favor brewed coffee.

Exercise 8.25

The desired information is

n = sample size = 594 + 1939 = 2533

$\hat{p}$ = the proportion of older adults who are pet owners = 594/2533 = 0.2345

C = level of confidence requested = 0.95

z^* = upper (1 – C)/2 critical value of the standard normal distribution
= upper 0.025 critical value of the standard normal distribution
= 1.96

Substituting these values into the formula for the confidence interval yields

$$\hat{p} \pm z^* \sqrt{\frac{\hat{p}(1-\hat{p})}{n}} = 0.2345 \pm 1.96\sqrt{\frac{0.2345(1-0.2345)}{2533}}$$

$$= 0.2345 \pm (1.96)(0.0084) = 0.2345 \pm 0.0165 = (0.218, 0.251)$$

Exercise 8.31

From the statement of the exercise we have

m = desired margin of error = 0.075

C = desired level of confidence = 0.95

z^* = the upper $(1 - C)/2$ critical value of the standard normal distribution

= 1.96

The required sample size n is thus

$$n = \left(\frac{z^*}{2m}\right)^2 = \left(\frac{1.96}{2(0.075)}\right)^2 = 170.738$$

which we round up to 171.

SECTION 8.2

OVERVIEW

Confidence intervals and tests designed to compare two population proportions are based on the **difference in the sample proportions** $D = \hat{p}_1 - \hat{p}_2$ where

$$\hat{p}_1 = \frac{X_1}{n_1} \quad \text{and} \quad \hat{p}_2 = \frac{X_2}{n_2}$$

and X_1 and X_2 are the number of successes in each group.

The formula for the level C confidence interval is

$$\hat{p}_1 - \hat{p}_2 \pm z^* \text{SE}_D$$

where z^* is the upper $(1 - C)/2$ standard normal critical value, SE_D is the **standard error for the difference in the two proportions** computed as

$$\text{SE}_D = \sqrt{\frac{\hat{p}_1(1-\hat{p}_1)}{n_1} + \frac{\hat{p}_2(1-\hat{p}_2)}{n_2}}$$

and $m = z^* \text{SE}_D$ is the **margin of error.** Use this interval for 90%, 95%, or 99% confidence when the number of successes and the number of failures in both samples are at least 10.

The **plus four estimate** of the difference in two population proportions is $\tilde{D} = \tilde{p}_1 - \tilde{p}_2$ where

$$\tilde{p}_1 = \frac{X_1 + 1}{n_1 + 2} \quad \text{and} \quad \tilde{p}_2 = \frac{X_2 + 1}{n_2 + 2}$$

The **standard error for the difference** $\tilde{D}$ is

$$\text{SE}_{\tilde{D}} = \sqrt{\frac{\tilde{p}_1(1-\tilde{p}_1)}{n_1+2} + \frac{\tilde{p}_2(1-\tilde{p}_2)}{n_2+2}}$$

and the **plus four level *C* confidence interval** is

$$\tilde{D} \pm m$$

with **margin of error** $m = z^* \, \text{SE}_{\tilde{D}}$.

Significance tests for the equality of the two proportions, H_0: $p_1 = p_2$, use a different standard error for the difference in the sample proportions that is based on a **pooled estimate** of the common (under H_0) value of p_1 and p_2,

$$\hat{p} = \frac{X_1 + X_2}{n_1 + n_2}$$

The test uses the *z statistic*

$$z = \frac{\hat{p}_1 - \hat{p}_2}{\text{SE}_{Dp}}$$

where

$$\text{SE}_{Dp} = \sqrt{\hat{p}(1-\hat{p})\left(\frac{1}{n_1} + \frac{1}{n_2}\right)}$$

and *P*-values are computed using Table A of the standard normal distribution.

GUIDED SOLUTIONS

Exercise 8.49

KEY CONCEPTS: Testing equality of two population proportions, confidence intervals for the difference between two population proportions

Let p_1 denote the proportion of all Internet users who downloaded music onto their computers two years before the 2005 survey, and p_2 denote the proportion of all Internet users who downloaded music onto their computers at the time of the 2005 survey. Is the alternative one- or two-sided? Read the wording of the problem carefully and state the hypotheses below.

H_0:
H_a:

To compute the test statistic, first we need to calculate $\hat{p}$, the pooled estimate of p_1 and p_2. Using the formulas in the text, you first need to find X_1 and X_2, the number of Internet users who downloaded music in the survey two years before 2005 and the number who downloaded music in the 2005 survey. The values of X_1 and X_2 can be computed from the sample size of the survey and the sample proportions $\hat{p}_1$ and $\hat{p}_2$. Make sure to round X_1 and X_2 to the nearest integer as they are counts. Compute

$X_1 =$

$X_2 =$

$$\hat{p} = \frac{X_1 + X_2}{n_1 + n_2} =$$

Use $\hat{p}$ to find the standard error of the difference in the proportions and then the value of the z statistic. Remember, in the z statistic $\hat{p}_1$ and $\hat{p}_2$ are the proportions who downloaded music in the survey conducted two years before 2005 and in the 2005 survey, respectively.

$$\text{SE}_{Dp} = \sqrt{\hat{p}(1-\hat{p})\left(\frac{1}{n_1} + \frac{1}{n_2}\right)} =$$

$$z = \frac{\hat{p}_1 - \hat{p}_2}{\text{SE}_{Dp}} =$$

What is the *P*-value and your conclusion?

The level C confidence interval for $p_1 - p_2$ is $\hat{p}_1 - \hat{p}_2 \pm z^* \text{SE}_D$, where

$$\text{SE}_D = \sqrt{\frac{\hat{p}_1(1-\hat{p}_1)}{n_1} + \frac{\hat{p}_2(1-\hat{p}_2)}{n_2}} =$$

To complete the calculations for the confidence interval, z^* is the upper $(1 - C)/2$ standard normal critical value and $\hat{p}_1$ and $\hat{p}_2$ are the proportions who downloaded music in the survey conducted two years before 2005 and in the 2005 survey, respectively. What is the numerical value of C? Use it to find z^*. Compute the confidence interval in the space below and explain what information is provided in the interval that is not given in the significance test results.

$\hat{p}_1 - \hat{p}_2 \pm z^* \text{SE}_D =$

Exercise 8.51

KEY CONCEPTS: Confidence intervals and tests for the difference between two population proportions

The data give the number of companies offering stock incentives to key employees from two groups: high-tech companies and non-high-tech companies. The data are reproduced below.

Group	n	X(offered incentive)
High-tech companies	91	73
Non-high-tech companies	109	75

We want a 95% confidence interval. The number of successes and the number of failures in both samples are at least 10. The conditions required to use the confidence interval based on the difference in the two sample proportions are satisfied. First, calculate the two sample proportions below.

$$\hat{p}_1 = \frac{X_1}{n_1} =$$

and

$$\hat{p}_2 = \frac{X_2}{n_2} =$$

The level C confidence interval for $p_1 - p_2$ is $\hat{p}_1 - \hat{p}_2 \pm z^* \text{SE}_D$, where $\hat{p}_1$ and $\hat{p}_2$ were computed above and

$$\text{SE}_D = \sqrt{\frac{\hat{p}_1(1-\hat{p}_1)}{n_1} + \frac{\hat{p}_2(1-\hat{p}_2)}{n_2}} =$$

To complete the calculations for the confidence interval, z^* is the upper $(1 - C)/2$ standard normal critical value. What is the numerical value of C? Use it to find z^*. Compute the confidence interval in the space below.

$$\hat{p}_1 - \hat{p}_2 \pm z^* \text{SE}_D =$$

(b) Let p_1 denote the proportion of high-tech companies that offer stock options to key employees and p_2 denote the proportion of non-high-tech companies that offer stock options to key employees. First, state the hypotheses in terms of p_1 and p_2.

H_0:
H_a:

The sample proportions $\hat{p}_1$ and $\hat{p}_2$ were computed in part (a). Now compute $\hat{p}$ (the pooled estimate of p_1 and p_2) and the standard error of the difference in proportions that uses this pooled estimate:

$$\hat{p} = \frac{X_1 + X_2}{n_1 + n_2} =$$

$$\text{SE}_{Dp} = \sqrt{\hat{p}(1-\hat{p})\left(\frac{1}{n_1}+\frac{1}{n_2}\right)} =$$

Finally, the value of the z statistic is

$$z = \frac{\hat{p}_1 - \hat{p}_2}{\text{SE}_{Dp}} =$$

and

P-value =

(c) Summarize your analysis and conclusions.

Complete Solutions

Exercise 8.49

The hypotheses are H_0: $p_1 = p_2$ and H_a: $p_1 \neq p_2$, since the problem says we are interested in whether or not there was a change. That is, the alternative is two-sided since we are interested in differences in either direction. The numbers who downloaded music in each year are $X_1 = n_1\hat{p}_1 = (1421)(0.29) = 412$ and $X_2 = n_2\hat{p}_2 = (1421)(0.22) = 313$. The pooled estimate $\hat{p}$ of p_1 and p_2 is

$$\hat{p} = \frac{X_1 + X_2}{n_1 + n_2} = \frac{412+313}{1421+1421} = \frac{725}{2842} = 0.255$$

which, in this case, is the average of the two sample proportions because the sample sizes are equal. The standard error of the difference in the two proportions and the z statistic are

$$\text{SE}_{Dp} = \sqrt{\hat{p}(1-\hat{p})\left(\frac{1}{n_1}+\frac{1}{n_2}\right)} = \sqrt{0.255(1-0.255)\left(\frac{1}{1421}+\frac{1}{1421}\right)} = 0.0164$$

$$z = \frac{\hat{p}_1 - \hat{p}_2}{\text{SE}_{Dp}} = \frac{0.29-0.22}{0.0164} = 4.27$$

since $\hat{p}_1 = 0.29$ and $\hat{p}_2 = 0.22$. Because the test is two-sided, the P-value is $2 \times P(Z > 4.29) \approx 0$, so there is very strong evidence of a difference in the proportions of Internet users who downloaded music onto their computers during the two survey years (more Internet users downloaded music before the filing of lawsuits by the recording industry).

To compute the confidence interval, the sample proportions $\hat{p}_1 = 0.29$ and $\hat{p}_2 = 0.22$ are given in the exercise. The standard error of their difference is

$$\text{SE}_D = \sqrt{\frac{\hat{p}_1(1-\hat{p}_1)}{n_1} + \frac{\hat{p}_2(1-\hat{p}_2)}{n_2}} = \sqrt{\frac{0.29(1-0.29)}{1421} + \frac{0.22(1-0.22)}{1421}} = 0.0163$$

For the 95% confidence interval, C = 0.95 and we need the upper $(1 - C)/2 = (1 - 0.95)/2 = 0.025$ standard normal critical value. This can be gotten most easily from the bottom row of Table D, with z^* = 1.96. The confidence interval is given by

$$\hat{p}_1 - \hat{p}_2 \pm z^* \text{SE}_D = (0.29 - 0.22) \pm (1.96)(0.0163) = 0.07 \pm 0.032$$

The difference in proportions is between 0.038 and 0.102. From the test, we know there is strong evidence of a difference in the proportion of Internet users downloading music in the two survey years. With 95% confidence, the difference could be as small as about 4% and as large as 10%. The confidence interval not only tells us that there is a difference, but provides information on the size of the difference. In this case the drop in the proportion of Internet users downloading music appears reasonably large given that the percent initially downloading music was only around 29%.

Exercise 8.51

(a) The sample proportions are

$$\hat{p}_1 = \frac{X_1}{n_1} = \frac{73}{91} = 0.8022$$

and

$$\hat{p}_2 = \frac{X_2}{n_2} = \frac{75}{109} = 0.6881$$

and the standard error of the difference is

$$\text{SE}_D = \sqrt{\frac{\hat{p}_1(1-\hat{p}_1)}{n_1} + \frac{\hat{p}_2(1-\hat{p}_2)}{n_2}} = \sqrt{\frac{0.8022(1-0.8022)}{91} + \frac{0.6881(1-0.6881)}{109}} = 0.0609$$

For the 95% confidence interval, C = 0.95 and we need the upper $(1 - C)/2 = (1 - 0.95)/2 = 0.025$ standard normal critical value. This can be gotten most easily from the bottom row of Table D, with z^* = 1.96. The confidence interval is given by

$$\hat{p}_1 - \hat{p}_2 \pm z^* \text{SE}_D = (0.8022 - 0.6881) \pm 1.96(0.0609) = 0.1141 \pm 0.1194$$

The difference in proportions is between –0.0053 and 0.2335.

(b) The question of interest is: Do high-tech companies tend to offer stock options more often than other companies? The hypotheses are H_0: $p_1 = p_2$ and H_a: $p_1 > p_2$. The sample proportions are $\hat{p}_1 = 0.8022$ and $\hat{p}_2 = 0.6681$. The pooled estimate of p_1 and p_2 is

$$\hat{p} = \frac{X_1 + X_2}{n_1 + n_2} = \frac{73 + 75}{91 + 109} = 0.74$$

and the standard error of the difference in the two proportions is

$$\text{SE}_{Dp} = \sqrt{\hat{p}(1-\hat{p})\left(\frac{1}{n_1} + \frac{1}{n_2}\right)} = \sqrt{0.74(1-0.74)\left(\frac{1}{91} + \frac{1}{109}\right)} = 0.0623$$

Thus, the test statistic for testing the hypotheses is

$$z = \frac{\hat{p}_1 - \hat{p}_2}{\text{SE}_{Dp}} = \frac{0.8022 - 0.6881}{0.0623} = 1.83$$

From Table A we find $P(Z > 1.83) = 0.0336$.

(c) With a P-value of 0.0336, there is strong evidence that high-tech companies tend to offer stock options more often than other companies. However, as we can see from the width of the confidence interval, the sample sizes are not large enough to give a precise interval. The 95% confidence interval on the difference goes from approximately 0% to 25% and doesn't provide very precise information about the size of the difference.

CHAPTER 9

ANALYSIS OF TWO-WAY TABLES

OVERVIEW

In this chapter, we discuss inference for two-way tables. There are two common models that generate data which can be summarized in a two-way table. In the first model, independent SRSs are drawn from c populations and each observation is classified according to a categorical variable that has r possible values. In the table, if the c populations form the columns and the categorical classification variable forms the rows then the null hypothesis is that the distribution of the row categorical variable is the same for each of the c populations. In the second model, an SRS is drawn from a single population, and the observations are cross-classified according to two categorical variables having r and c possible values, respectively. In this model the null hypothesis is that the row and column variables are independent. When one of the variables is an explanatory variable and the other is a response, the explanatory variable is often used to form the columns of the table while the response forms the rows.

A test of the null hypothesis is carried out using X^2 in both models. Although the data are generated differently, the question in both cases is quite similar: are the distributions of the column variables the same? The cell counts are compared to the **expected cell counts** under the null hypothesis. The expected cell counts are computed using the formula

$$\text{expected count} = \frac{\text{row total} \times \text{column total}}{n}$$

where n is the total number of observations.

The **chi-square statistic** is used to test the null hypothesis by comparing the observed counts with the expected counts:

$$X^2 = \sum \frac{(\text{observed count} - \text{expected count})^2}{\text{expected count}}$$

When the null hypothesis is true, the distribution of X^2 is approximately χ^2 with $(r-1)(c-1)$ degrees of freedom. The P-value is the probability of getting differences between observed and expected counts as large as we did and is computed as $P(\chi^2 \geq X^2)$. The use of the χ^2 distribution is an approximation that works well when the average of the expected counts exceeds 5 and all of the individual expected counts are greater than 1. In the special case of the 2×2 table, all expected counts should exceed 5 before applying the approximation.

The comparison of the proportion of "successes" in two populations described in Chapter 8 leads to a 2 × 2 table. The chi-square test and the two-sample z test from Section 8.2 always give exactly the same result because the X^2 statistic is equal to the square of the z statistic and the $\chi^2(1)$ critical values are equal to the squares of the corresponding $N(0, 1)$ critical values. The advantage of the z test is that it can be used for either one-sided or two-sided alternatives.

In addition to computing the X^2 statistic, tables or bar graphs should be examined to describe the relationship between the two variables.

Data for n observations on a categorical variable with k possible outcomes can be summarized as observed counts $n_1, n_2, \cdots, n_k$ in the k cells. The **chi-square goodness of fit test** can test the null hypothesis that specifies probabilities $p_1, p_2, \cdots, p_k$ for the possible outcomes.

The expected counts for each cell are obtained by multiplying the total number of observations n by the specified probability:

$$\text{expected count} = np_i$$

The **chi-square statistic** is a measure of how much the observed and expected counts differ and is computed using the formula

$$X^2 = \sum \frac{(\text{observed count} - \text{expected count})^2}{\text{expected count}}$$

The degrees of freedom are $k - 1$ and the P-values are computed from the appropriate chi-square distribution.

GUIDED SOLUTIONS

Exercise 9.7

KEY CONCEPTS: Joint, marginal and conditional distributions in two-way tables

(a) For convenience, the table given in the problem is reproduced below. The entries are in thousands of U.S. college students. We have filled in the totals for you.

U.S. college students by age and status

	Status		
Age	Full-time	Part-time	Total
15 - 19	3553	329	3882
20 - 24	5710	1215	6925
25 - 34	1825	1864	3689
35 and over	901	1983	2884
Total	11989	5391	17380

The joint distribution is the collection of cell proportions for the two categorical variables. A cell proportion is the proportion that each cell count is of the total number of counts in the table (in this case, 17,380 thousands of persons). For example, the cell proportion corresponding to the age group "15 - 19" and "Full-time" status is 3553/17,380 = 0.204.

To answer the question, you need to compute each cell proportion and enter them in the table below. This will be the joint distribution. You can double-check your work by verifying that the sum of all the proportions is 1.

U.S. college students by age and status

Age	Status: Full-time	Part-time
15 - 19	0.204	
20 - 24		
25 - 34		
35 and over		

(b) The marginal distribution of age can be found from the totals that were given to you for each age group in the rightmost column of the table. Each value in this column must be divided by the total number of counts, namely 17,380, to give the proportion in each age group. Compute each of these proportions and enter the results in the table below.

Age group	Proportion
15 - 19	
20 - 24	
25 - 34	
35 and over	

Now display the results of this table graphically on the axes provided. The bar for 25 to 34 is provided.

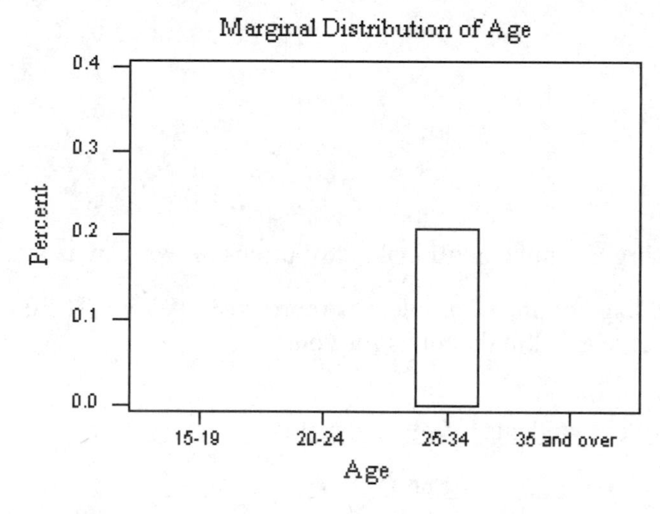

(c) The marginal distribution of status can be found from the totals for full-time and part-time status given in the bottom row of the table. Each value in this row must be divided by the total number of counts, namely 17,380, to give the proportion in each status. Compute each of these proportions and enter the results in the table below.

Status	Proportion
Full-time	
Part-time	

Now display the results of this table graphically on the axes provided.

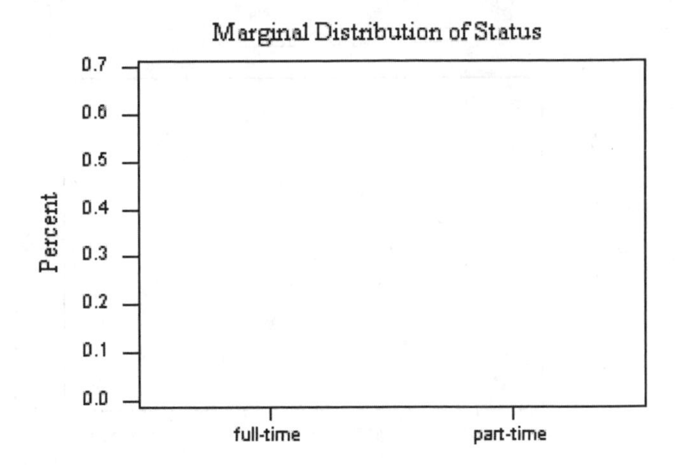

(d) To compute the conditional distribution of age for a particular status category one must divide each cell entry in the row corresponding to the age by the row total (entry in bottom row of the table). For example, for the status "full-time" the conditional distribution involves dividing the entries 3553, 5710, 1825 and 901 each by 11,989.

Carry out these calculations and enter the results in the table below.

	Status	
Age	Full-time	Part-time
15 - 19		
20 - 24		
25 - 34		
35 and over		

To summarize these results graphically, make a separate bar graph for each status category. To assist you, we have provided the axes for the two plots below.

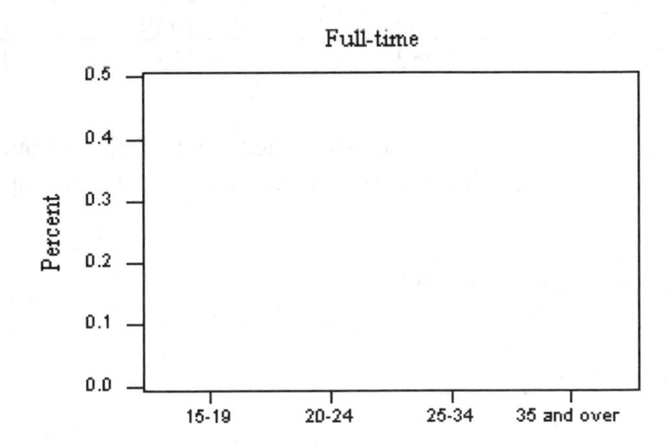

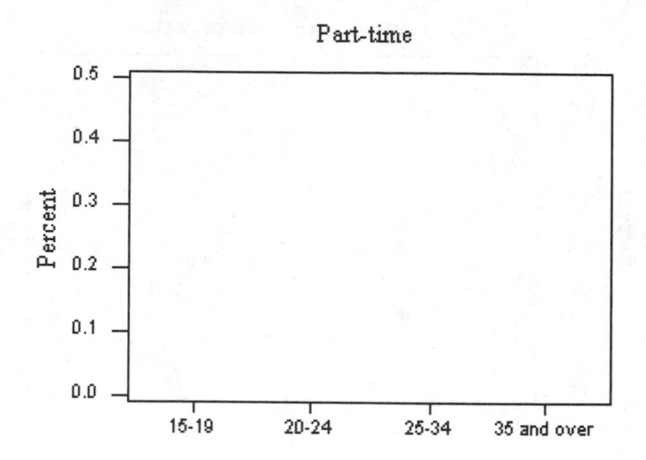

(e) How do the distributions differ?

Exercise 9.17

KEY CONCEPTS: Testing that the distributions of the response variable are the same in *c* populations

(a) The data are reproduced below with the row and column totals included.

Missed a class	Nonbinger	Occasional binger	Frequent binger	Total
No	4617	2047	1176	7840
Yes	446	915	1959	3320
Total	5063	2962	3135	11160

The comparison of the proportions of students missing each type of class would be of most interest in examining the relationship between missing a class and drinking status. Compute these proportions to complete the table below.

Status	Proportion
Nonbinger	
Occasional binger	
Frequent binger	

(b) The marginal distribution of drinking status can be found from the totals for each type of drinking status given in the bottom row of the table. Each value in this row must be divided by the total number of students, namely 11,160, to give the proportion in each status. Compute each of these proportions and display the results graphically on the following page.

(c) The relative risk of missing a class for occasional bingers versus nonbingers is the proportion of students who missed class in the occasional binger group divided by the proportion who missed class in the nonbinger group. Compute the relative risk below.

Relative risk =

The relative risk of missing a class for frequent bingers versus nonbingers is the proportion of students who missed class in the frequent binger group divided by the proportion who missed class in the nonbinger group. Compute the relative risk below.

Relative risk =

Summarize what these two relative risks tell you about drinking status and missing classes.

(d) To test the hypothesis that the probability of missing a class is the same for frequent, occasional, and nonbingers, we must compute the X^2 statistic,

$$X^2 = \sum \frac{(\text{observed count} - \text{expected count})^2}{\text{expected count}}$$

The observed values for each cell can be read from the table. The expected counts are computed for each cell in the table using the formula

$$\text{expected count} = \frac{\text{row total} \times \text{column total}}{n}$$

where n is the total number of individuals in the table. Compute the expected counts and enter the values in the space below each of the observed counts in the following table.

Missed a class	Nonbinger	Occasional binger	Freguent binger	Total
No	4617	2047	1176	7840
Yes	446	915	1959	3320
Total	5063	2962	3135	11160

Now compute

$$X^2 = \sum \frac{(\text{observed count} - \text{expected count})^2}{\text{expected count}} =$$

What is the *P*-value? What do you conclude?

Exercise 9.30

KEY CONCEPTS: 2 × 2 table, chi-square test, two-sample *z* test

(a) Let p_1 denote the proportion of women customers in City 1 and p_2 the proportion of women customers in City 2. First, state the hypotheses in terms of p_1 and p_2.

H_0:
H_a:

From the problem, there were $n_1 = 241$ customers in City 1, and $X_1 = 203$ were women. There were $n_2 = 218$ customers in City 2, and $X_2 = 150$ were women. First, find the sample proportion of women in both cities.

$\hat{p}_1 = X_1 / n_1 =$

$\hat{p}_2 = X_2 / n_2 =$

To compute the denominator of the *z* statistic, you must compute compute $\hat{p}$ (the pooled estimate of p_1 and p_2) and the standard error of the difference in proportions that uses this pooled estimate:

$$\hat{p} = \frac{X_1 + X_2}{n_1 + n_2} =$$

$$\text{SE}_{Dp} = \sqrt{\hat{p}(1-\hat{p})\left(\frac{1}{n_1} + \frac{1}{n_2}\right)} =$$

Finally, the value of the *z* statistic is

$$z = \frac{\hat{p}_1 - \hat{p}_2}{\text{SE}_{Dp}} =$$

What is the *P*-value?

(b) The output that follows is from Minitab. If you don't have access to software, you may want to review Exercise 9.17 for the details of doing the calculations by hand.

Chi-Square Test: City 1, City 2

```
Expected counts are printed below observed counts
Chi-Square contributions are printed below expected counts

          City 1  City 2  Total
    Men       38      68    106
           55.66   50.34
           5.601   6.192

    Women    203     150    353
          185.34  167.66
           1.682   1.859

Total     241       218     459

Chi-Sq = 15.334, DF = 1, P-Value = 0.000
```

The value of $X^2 = 15.334$. Verify that this is the square of the z statistic from part (a) and show the P-value agrees with part (a) using Table F.

(c) We want a 95% confidence interval. The number of successes and the number of failures in both samples are at least 10. The conditions required to use the confidence interval based on the difference in the two sample proportions are satisfied.

The level C confidence interval for $p_1 - p_2$ is $\hat{p}_1 - \hat{p}_2 \pm z^*\text{SE}_D$, where $\hat{p}_1$ and $\hat{p}_2$ were computed in part (a) and

$$\text{SE}_D = \sqrt{\frac{\hat{p}_1(1-\hat{p}_1)}{n_1} + \frac{\hat{p}_2(1-\hat{p}_2)}{n_2}} =$$

To complete the calculations for the confidence interval, z^* is the upper $(1 - C)/2$ standard normal critical value. What is the numerical value of C? Use it to find z^*. Compute the confidence interval in the space below.

$\hat{p}_1 - \hat{p}_2 \pm z^*\text{SE}_D =$

Exercise 9.31

KEY CONCEPTS: Chi-square tests, degrees of freedom, *P*-values

When the null hypothesis is true, the distribution of X^2 is approximately χ^2 with $(r-1)(c-1)$ degrees of freedom. What are the values of r and c in this problem and the associated degrees of freedom for chi-square?

r =
c =
df =

The P-value is the probability of getting differences between observed and expected counts as large as we did and is computed as $P(\chi^2 \geq X^2)$. The value of X^2 is given as 3.995. Using Table F, what can you say about the P-value? If you have software available, compute the P-value using your software.

P-value using Table F:

P-value using software:

Is there good evidence that customers at the two stores have different income distributions?

COMPLETE SOLUTIONS

Exercise 9.7

(a) The joint distribution is as indicated in the completed table below.

U.S. College students by age and status

	Status	
Age	Full-time	Part-time
15 - 19	0.204	0.019
20 - 24	0.329	0.070
25 - 34	0.105	0.107
35 and over	0.052	0.114

(b) The marginal distribution of age and graphical display are given below.

	Age group			
	15 to 19	20 to 24	25 to 34	35 and over
Proportion	0.223	0.398	0.212	0.166

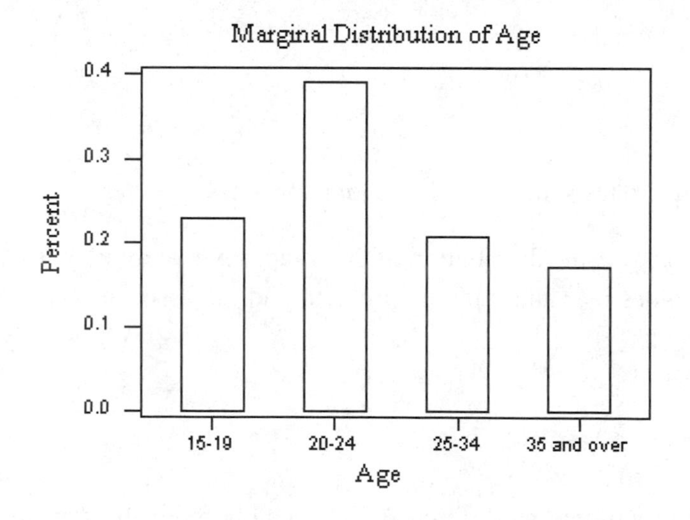

(c) The marginal distribution of status and graphical display are given below.

Status	Proportion
Full-time	0.690
Part-time	0.310

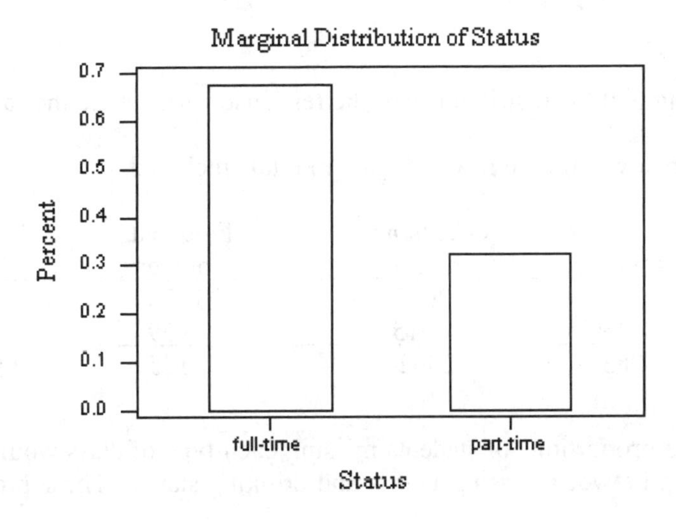

The conditional distributions of age for full-time and part-time is summarized in the columns of the table below.

	Status	
Age	Full-time	Part-time
15 - 19	0.296	0.061
20 - 24	0.476	0.225
25 - 34	0.152	0.346
35 and over	0.075	0.368

Graphical displays of these conditional distributions follow.

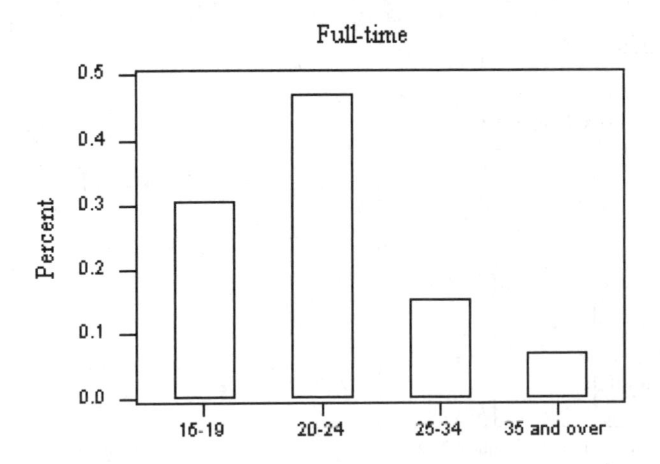

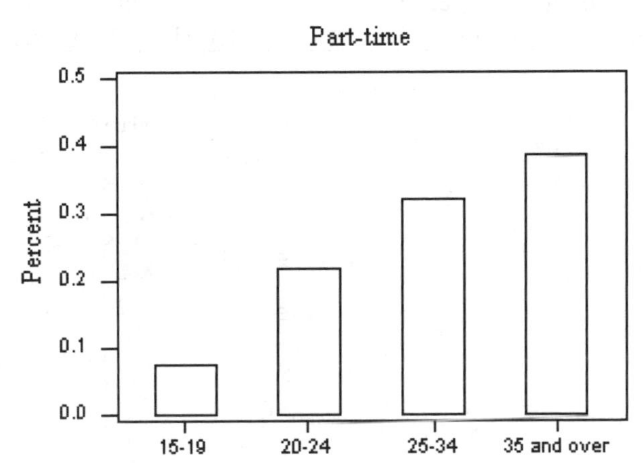

The age distributions of full-time and part-time students are quite different. The proportions in each age category for part-time increase with age, with the majority being in the 35 and over category. For the full-time students most are in the 15 to 19 and the 20 to 24 age categories. This is not surprising, as many part-time students are working and have families and are older than the full-time students, who have generally gone directly from high school to college.

Exercise 9.17

KEY CONCEPTS: Testing that the distributions of the response variable are the same in *c* populations

The data are reproduced below with the row and column totals included.

Missed a class	Nonbinger	Occasional binger	Frequent binger	Total
No	4617	2047	1176	7840
Yes	446	915	1959	3320
Total	5063	2962	3135	11160

(a) The comparison of the proportions of students missing each type of class would be of most interest in examining the relationship between missing a class and drinking status. These proportions are computed in the table below.

Status	Proportion who missed a class
Nonbinger	446/5063 = 0.088
Occasional binger	915/2962 = 0.309
Frequent binger	1959/3135 = 0.625

The pattern is clear. As binge drinking increases from nonbinger to occasional binger to frequent binger, the proportion of students missing a class increases.

(b) The marginal distribution of drinking status is given below with the bar chart displaying this distribution.

Status	Proportion
Nonbinger	5063/11160 = 0.454
Occasional binger	2962/11160 = 0.265
Frequent binger	3135/11160 = 0.281

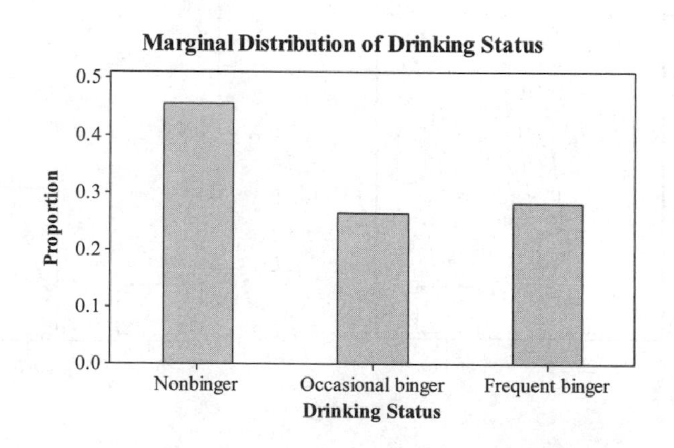

(c) The relative risk of missing a class for occasional bingers versus nonbingers is can be found using the proportions that were computed in part (a).

Relative risk = 0.309/0.088 = 3.51

The relative risk of missing a class for frequent bingers versus nonbingers is

Relative risk = 0.625/0.088 = 7.10

Compared to nonbingers, occasional bingers are about 3.5 times more likely to miss class because of drinking and frequent bingers about 7 times more likely.

(d) To test the hypothesis that the probability of missing a class is the same for frequent, occasional and nonbingers, we must compute the X^2 statistic,

$$X^2 = \sum \frac{(\text{observed count} - \text{expected count})^2}{\text{expected count}}$$

The observed values for each cell can be read from the table. The expected counts are computed for each cell in the table using the formula

$$\text{expected count} = \frac{\text{row total} \times \text{column total}}{n}$$

where n is the total number of individuals in the table. The values of the expected counts are entered below each of the observed counts in the following table.

Missed a class	Nonbinger	Occasional binger	Freguent binger	Total
No	4617	2047	1176	7840
	3556.80	2080.83	2202.37	
Yes	446	915	1959	3320
	1506.20	881.17	932.63	
Total	5063	2962	3135	11160

The details of computing the expected counts are as follows.

$n = 11160$

(No, nonbinger): $\text{expected count} = \dfrac{7840 \times 5063}{11160} = 3556.80$

(Yes, nonbinger): $\text{expected count} = \dfrac{3320 \times 5063}{11160} = 1506.20$

(No, occasional binger): $\text{expected count} = \dfrac{7840 \times 2962}{11160} = 2080.83$

(Yes, occasional binger): $\text{expected count} = \dfrac{3320 \times 2962}{11160} = 881.17$

(No, frequent binger): $\text{expected count} = \dfrac{7840 \times 3135}{11160} = 2202.37$

(Yes, frequent binger): expected count = $\dfrac{3320 \times 3135}{11160} = 932.63$

We now compute

$$X^2 = \sum \frac{(\text{observed count} - \text{expected count})^2}{\text{expected count}} = \frac{(4617 - 3556.80)^2}{3556.80} + \frac{(2047 - 2080.83)^2}{2080.83}$$

$$+ \frac{(1176 - 2202.37)^2}{2202.37} + \frac{(446 - 1506.20)^2}{1506.20}$$

$$+ \frac{(915 - 881.17)^2}{881.17} + \frac{(1959 - 932.63)^2}{932.63}$$

$$= 316.019 + 0.550 + 478.316 + 746.262 + 1.299 + 1129.517$$

$$= 2671.963$$

$X^2 = 2671.963$ and has $(r - 1)(c - 1) = (2 - 1)(3 - 1) = 2$ degrees of freedom. Going to Table F and using the line corresponding to 2 degrees of freedom, we find this exceeds the largest entry of 15.20, corresponding to an upper tail area of 0.0005. We conclude that the *P*-value = $P(\chi^2 \geq 2671.963)$ is less than 0.0005. Thus, there is very strong evidence against the null hypothesis that the probability of missing class is the same for the three statuses of drinking. The probability goes up as binge drinking increases.

Exercise 9.30

(a) We simply test to see if the proportions are different. The hypotheses are H_0: $p_1 = p_2$ and H_a: $p_1 \neq p_2$. The sample proportions are $\hat{p}_1 = X_1 / n_1 = 203/241 = 0.8423$ and $\hat{p}_2 = X_2 / n_2 = 150/218 = 0.6881$. The pooled estimate of p_1 and p_2 is

$$\hat{p} = \frac{X_1 + X_2}{n_1 + n_2} = \frac{203 + 150}{241 + 218} = 0.7691$$

and the standard error of the difference in the two proportions is

$$\text{SE}_{Dp} = \sqrt{\hat{p}(1-\hat{p})\left(\frac{1}{n_1} + \frac{1}{n_2}\right)} = \sqrt{0.7691(1-0.7691)\left(\frac{1}{241} + \frac{1}{218}\right)} = 0.0394$$

Thus, the test statistic for testing the hypotheses is

$$z = \frac{\hat{p}_1 - \hat{p}_2}{\text{SE}_{Dp}} = \frac{0.8423 - 0.6881}{0.0394} = 3.914$$

From Table A we find the *P*-value is $2 \times P(Z > 3.914) \approx 0$.

(b) The computer output for the chi-square is given in the Guided Solutions. The value of $X^2 = 15.334$. The square of the z statistic is $(3.914)^2 = 15.32$, which agrees with the value of X^2 except for rounding error. Using Table F with $(r-1)(c-1) = 1$ degrees of freedom, we see the *P*-value is less than 0.0005 since the X^2 statistic value of 15.334 is larger than the largest $(\chi^2)^*$ value in the table.

(c) From part (a) $\hat{p}_1 = 0.8423$ and $\hat{p}_2 = 0.688$, giving

$$\text{SE}_D = \sqrt{\frac{\hat{p}_1(1-\hat{p}_1)}{n_1} + \frac{\hat{p}_2(1-\hat{p}_2)}{n_2}} = \sqrt{\frac{0.8423(1-0.8423)}{241} + \frac{0.6881(1-0.6881)}{218}} = 0.0392$$

To complete the calculations for the confidence interval, $z^* = 1.96$ is the upper $(1-C)/2 = 0.025$ standard normal critical value, giving the confidence interval

$$\hat{p}_1 - \hat{p}_2 \pm z^*\text{SE}_D = (0.8423 - 0.6881) \pm 1.96(0.0392) = 0.1542 \pm 0.0768 = (0.0774, 0.2310).$$

Exercise 9.31

Because $r = 5$ and $c = 2$, the degrees of freedom are $(5-1)(2-1) = 4$. According to the row corresponding to df = 4 in Table F, the value 5.39 corresponds to a tail probability of 0.25. So the *P*-value exceeds 0.25. Statistical software gives $P(\chi^2 \geq 3.955) = 0.4121$ as the *P*-value. There is no evidence of a difference in the income distributions of the customers at the two stores.

CHAPTER 10

INFERENCE FOR REGRESSION

OVERVIEW

The statistical model for **simple linear regression** is

$$y_i = \beta_0 + \beta_1 x_i + \varepsilon_i$$

where i = 1, 2, ⋯, n. The deviations ε_i are assumed to be independent and normally distributed with mean 0 and standard deviation σ. The **parameters** of the model are the intercept β_0, the slope β_1, and σ. The parameters β_0 and β_1 are estimated by the slope b_0 and intercept b_1 of the **least-squares regression line.** Given n observations on an explanatory variable x and a response variable y,

$$(x_1, y_1), (x_2, y_2), \cdots, (x_n, y_n)$$

recall that the formulas for the slope and intercept of the least-squares regression line are

$$b_1 = r\frac{s_y}{s_x}$$

and

$$b_0 = \bar{y} - b_1\bar{x}$$

where r is the correlation between y and x, $\bar{y}$ is the mean of the y observations, s_y is the standard deviation of the y observations, $\bar{x}$ is the mean of the x observations, and s_x is the standard deviation of the x observations. The standard deviation σ is estimated by

$$s = \sqrt{\frac{\sum e_i^2}{n-2}}$$

where the e_i are the **residuals**

$$e_i = y_i - \hat{y}_i$$

and

$$\hat{y}_i = b_0 + b_1 x_i$$

b_0, b_1, and s are usually calculated using a calculator or statistical software.

A **level *C* confidence interval** for β_1 is

$$b_1 \pm t^* \mathrm{SE}_{b_1}$$

where t^* is the upper $(1 - C)/2$ critical value for the $t(n - 2)$ distribution and

$$\mathrm{SE}_{b_1} = \frac{s}{\sqrt{\sum (x_i - \bar{x})^2}}$$

is the standard error of the slope b_1. SE_{b_1} is usually computed using a calculator or statistical software. (The formula above is actually given in Section 10.2 of the text but is reproduced here for continuity of exposition. If your class is not covering Section 10.2, you need not worry about the formula.)

The **test of the hypothesis** H_0: $\beta_1 = 0$ is based on the t statistic

$$t = \frac{b_1}{\mathrm{SE}_{b_1}}$$

with P-values computed from the $t(n - 2)$ distribution. There are similar formulas for confidence intervals and tests for β_0, but using the standard error of the intercept b_0

$$\mathrm{SE}_{b_0} = s\sqrt{\frac{1}{n} + \frac{\bar{x}^2}{\sum (x_i - \bar{x})^2}}$$

in place of SE_{b_1}. As with SE_{b_1}, SE_{b_0} is usually computed using a calculator or statistical software. (The formula above is actually given in Section 10.2 of the text but is reproduced here for continuity of exposition. If your class is not covering Section 10.2, you need not worry about the formula.) Note that inferences for the intercept is meaningful only in special cases.

The **estimated mean response** for the subpopulation corresponding to the value x^* of the explanatory variable is

$$\hat{\mu}_y = b_0 + b_1 x^*$$

A **level *C* confidence interval for the mean response** is

$$\hat{\mu}_y \pm t^* \mathrm{SE}_{\hat{\mu}}$$

where t^* is the upper $(1 - C)/2$ critical value for the $t(n - 2)$ distribution and

$$\mathrm{SE}_{\hat{\mu}} = s\sqrt{\frac{1}{n} + \frac{(x^* - \bar{x})^2}{\sum (x_i - \bar{x})^2}}$$

$\mathrm{SE}_{\hat{\mu}}$ is usually computed using a calculator or statistical software. (The formula above is actually given in Section 10.2 of the text but is reproduced here for continuity of exposition. If your class is not covering Section 10.2, you need not worry about the formula.)

The **estimated value of the response variable** y for a future observation from the subpopulation corresponding to the value x^* of the explanatory variable is

$$\hat{y} = b_0 + b_1 x^*$$

A **level *C* prediction interval** for the estimated response is

$$\hat{y} \pm t^* \mathrm{SE}_{\hat{y}}$$

where t^* is the upper $(1 - C)/2$ critical value for the $t(n - 2)$ distribution and

$$\mathrm{SE}_{\hat{y}} = s\sqrt{1 + \frac{1}{n} + \frac{(x^* - \bar{x})^2}{\sum(x_i - \bar{x})^2}}$$

$\mathrm{SE}_{\hat{y}}$ is usually computed using a calculator or statistical software. (The formula above is actually given in Section 10.2 of the text but is reproduced here for continuity of exposition. If your class is not covering Section 10.2, you need not worry about the formula.)

The **ANOVA (analyis of variance) table** for a linear regression gives the total sum of squares SSM for the model, the total sum of squares SSE for error, the total sum of squares SST for all sources of variation, the degrees of freedom DFM, DFE, and DFT for these sums of squares, the mean square MSM for the model, and the mean square MSE for error. The formulas for the sums of squares are

$$\mathrm{SSM} = \sum(\hat{y}_i - \bar{y})^2$$

$$\mathrm{SSE} = \sum(y_i - \hat{y}_i)^2$$

$$\mathrm{SST} = \sum(y_i - \bar{y})^2$$

and the degrees of freedom are

$$\mathrm{DFM} = 1$$

$$\mathrm{DFE} = n - 2$$

$$\mathrm{DFT} = n - 1$$

The mean sum of squares MS is defined by the relation

$$\mathrm{MS} = \frac{\text{sum of squares}}{\text{degrees of freedom}}$$

The ANOVA table usually has a form like the following.

Source	Degrees of freedom	Sum of squares	Mean Square	*F*
Model	DFM	SSM	MSM = SSM/DFM	MSM/MSE
Error	DFE	SSE	MSE = SSE/DFE	
Total	DFT	SST		

The **ANOVA *F*-statistic** is the ratio MSM/MSE and is used to test H_0: $\beta_1 = 0$ versus the two-sided alternative. Under H_0, this statistic has an $F(1, n - 2)$ distribution, which can be used to compute *P*-values using Table E.

When the variables y and x are jointly normal, the sample correlation is an estimate of the population correlation ρ. The test of H_0: $\rho = 0$ is based on the ***t*-statistic**

$$t = \frac{r\sqrt{n-2}}{\sqrt{1-r^2}}$$

which has a $t(n-2)$ distribution under H_0. This test statistic is numerically identical to the t statistic used to test H_0: $\beta_1 = 0$.

The **square of the sample correlation** can be written as

$$r^2 = \text{SSM/SST}$$

and is interpreted as the proportion of the variability in the response variable y that is explained by the explanatory variable x in the simple linear regression.

GUIDED SOLUTIONS

Exercise 10.35

KEY CONCEPTS: Scatterplots, least-squares equation, r^2, confidence intervals for the slope

(a) Draw your scatterplot in the axes provided below. What is the explanatory variable and what is the response variable here? Remember that in the scatterplot, the horizontal axis represents the explanatory variable and the vertical axis the response variable. Label your axes.

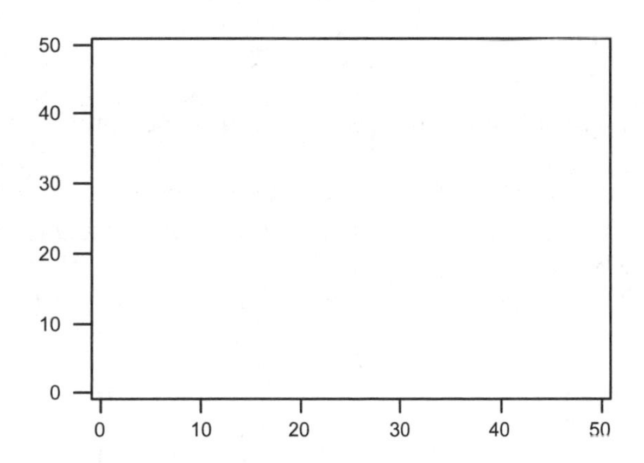

(b) Describe the form, strength, and direction of the relationship between HAV and MA. Are there any outliers or unusual observations in the plot?

(c) Give a statistical model that provides a framework for asking questions of interest in this problem.

(d) The researchers speculated that there is a positive association—more serious MA is associated with more serious HAV. Translate this into appropriate null and alternative hypotheses.

H_0: H_a:

(e) The data were entered into Minitab and a portion of the output is given below.

Regression Analysis: HAV versus MA

```
The regression equation is
HAV = 19.7 + 0.339 MA

Predictor        Coef    SE Coef        T       P
Constant       19.723      3.217     6.13   0.000
MA             0.3388     0.1782     1.90   0.065

S = 7.224       R-Sq = 9.1%       R-Sq(adj) = 6.6%
```

The test that is automatically included as part of the output is a test of H_0: $\beta_1 = 0$ and H_a: $\beta_1 \neq 0$. In particular, the P-value of 0.065 = 2 × $P(T \geq 1.90)$. In this case we have a one-sided alternative, so the P-value should be $P(T \geq 1.90)$, or half the value printed in the output. Give a short description of the results in the space below.

Exercise 10.39

KEY CONCEPTS: Scatterplots, least-squares equation, r^2, confidence intervals for the slope

(a) What is the explanatory variable and what is the response variable here? Remember that in the scatterplot, the horizontal axis represents the explanatory variable and the vertical axis the response variable. Draw your scatterplot in the axes provided below. Are there any outliers or unusual observations in the plot? Does the trend in lean over time appear to be linear?

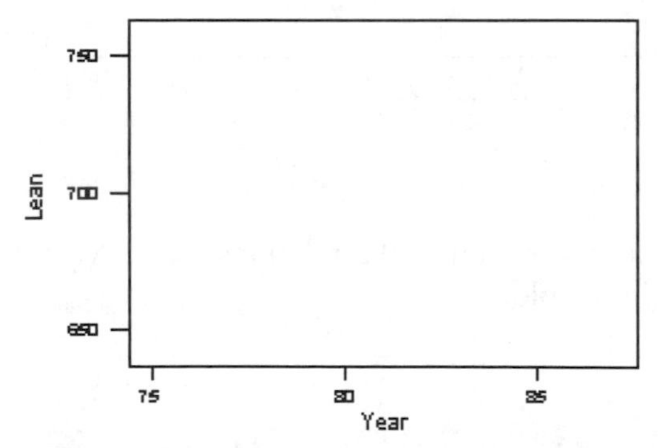

(b) If possible, use statistical software to compute the equation of the least-squares line and to compute the percent of variation in lean that is explained by this line. To help you answer these questions, the data were entered into Minitab and a portion of the output is provided below. If you use other software, the output should look quite similar.

Regression Analysis: Lean versus Year

```
The regression equation is
Lean = - 61.1 + 9.32 Year

Predictor       Coef    SE Coef        T        P
Constant      -61.12      25.13    -2.43    0.033
Year          9.3187     0.3099    30.07    0.000

S = 4.181       R-Sq = 98.8%     R-Sq(adj) = 98.7%

Analysis of Variance

Source            DF        SS        MS        F        P
Regression         1     15804     15804   904.12    0.000
Residual Error    11       192        17
Total             12     15997
```

Equation of the least-squares line:

Percent of variation in lean explained by the line:

(c) Which parameter measures the average rate of change of the lean? The confidence interval is not given directly on the Minitab output, but the basic quantities you need are provided. The formula for the confidence interval is

$$b_1 \pm t^* \mathrm{SE}_{b_1}$$

where t^* is the upper $(1 - C)/2$ critical value for the $t(n - 2)$ distribution and

$$\mathrm{SE}_{b_1} = \frac{s}{\sqrt{\sum (x_i - \bar{x})^2}}$$

is the standard error of the slope b_1. The values of b_1 and SE_{b_1} are contained in the Minitab output. You will need to use Table D to find t^*. Complete the calculations below.

$C =$
$t^* =$
$b_1 =$
$\mathrm{SE}_{b_1} =$

$b_1 \pm t^* \mathrm{SE}_{b_1} =$

Exercise 10.40

KEY CONCEPTS: Predictions, extrapolation

(a) The equation of the least-squares line was found in part (b) of Exercise 10.39. Use it to predict the lean in 1918, which has a value of $x^* = 18$.

$\hat{y} = b_0 + b_1 x^* =$

(b) The scatterplot below is of the value of lean versus year for the year 1918 and for the years 1975 through 1987. In addition, the equation of the least-squares line is drawn. You can use this to help answer the question posed in the exercise.

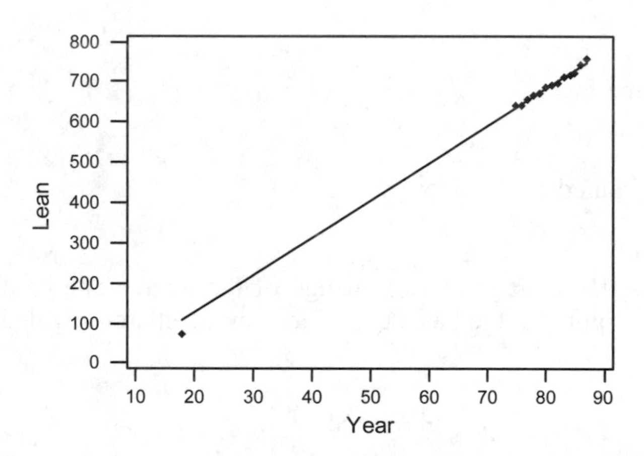

Exercise 10.41

KEY CONCEPTS: Prediction, prediction intervals, and confidence intervals

(a) The explanatory variable is the number of years since 1900. How would you code the year 2009?

(b) Use the coded value of 2009 and the equation of the least-squares line to predict the lean in 2009.

$\hat{y} = b_0 + b_1 x^* =$

(c) If you were giving a margin of error, should you use a confidence interval for a mean response or a prediction interval? Explain.

Exercise 10.47

KEY CONCEPTS: Analysis of variance table

Complete the analysis of variance table below by filling in the "Residual Error" row.

```
Analysis of Variance

Source            DF          SS          MS        F        P
Regression         1      3445.9      3445.9     9.50    0.005
Residual Error
Total             29     13598.3
```

Exercise 10.48

KEY CONCEPTS: Regression standard error and r^2

Both the regression standard error and r^2 can be computed from the entries in the completed analysis of variance table in Exercise 10.47. The required formulas are contained in Section 10.2 of your text.

$s = \sqrt{\text{MSE}} =$

$r^2 = \dfrac{\text{SSM}}{\text{SST}} =$

Exercise 10.49

KEY CONCEPTS: Standard error and confidence interval for the least-squares slope

The formula for the standard error of the slope is given in Section 10.2 as

$$\text{SE}_{b_1} = \frac{s}{\sqrt{\sum(x_i - \bar{x})^2}}$$

and requires $\sqrt{\sum(x_i - \bar{x})^2}$ and the regression standard error s. The regression standard error was computed in Exercise 10.48. We are given that the standard deviation of the S&P 500 returns for these years is 16.45%. Since S&P 500 returns is the explanatory variable x, this says that

$$\sqrt{\sum(x_i - \bar{x})^2 / n-1} = 16.45$$

You can use this relationship to find

$\sqrt{\sum(x_i - \bar{x})^2} =$

and from this

$$\mathrm{SE}_{b_1} = \frac{s}{\sqrt{\sum (x_i - \bar{x})^2}} =$$

The confidence interval for the slope is

$$b_1 \pm t^* \mathrm{SE}_{b_1}$$

where t^* is the upper $(1 - C)/2$ critical value for the $t(n - 2)$ distribution. The value of b_1 is contained in the regression equation that is part of the Minitab output in Exercise 10.47 in the text. The value of t^* can be found using Table D, and you have just calculated SE_{b_1}. Thus, the 95% confidence interval is

$b_1 \pm t^* \mathrm{SE}_{b_1} =$

Complete Solutions

Exercise 10.35

(a) The scatterplot below was produced by Minitab.

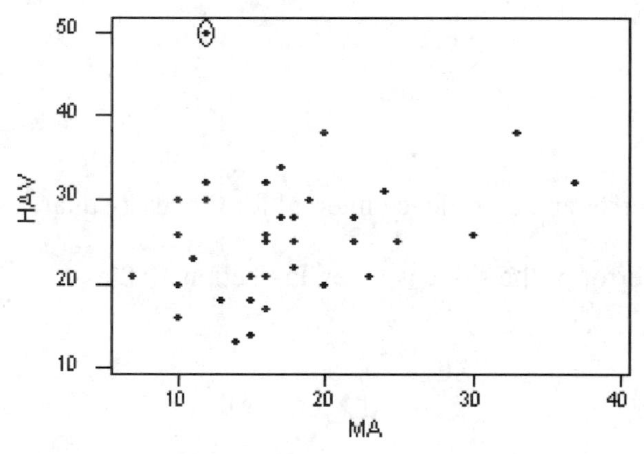

(b) The relationship between MA and HAV is positive, but fairly weak. A line would be a reasonable approximation to the relationship. There is one outlier, with an MA value of around 12, and it is circled in the scatterplot.

(c) An appropriate model that provides a framework for asking the question of interest in this problem is

$$\mathrm{HAV}_i = \beta_0 + \beta_1 \mathrm{MA}_i + \varepsilon_i \qquad \varepsilon_i \text{ are } N(0, \sigma)$$

(d) The researchers speculated that there is a positive association—more serious MA is associated with more serious HAV. Under the model in (c), a positive association corresponds to $\beta_1 > 0$. Thus, the alternative is one-sided. The hypotheses are

$$H_0\colon \beta_1 = 0 \text{ and } H_a\colon \beta_1 > 0$$

(e) The value of the test statistic for H_0: $\beta_1 = 0$ and H_a: $\beta_1 > 0$ is $t = 1.90$, with 36 degrees of freedom. The P-value is 0.065/2 = 0.033, which indicates a positive association between HAV and MA, although the relationship appears weak, with an r^2 value of only 9.1%. That is, MA by itself is not a very good predictor of HAV.

Exercise 10.39

(a) Since we are interested in whether year can be used to predict lean, we treat year as the explanatory variable and lean as the response. A scatterplot of year and lean is given below.

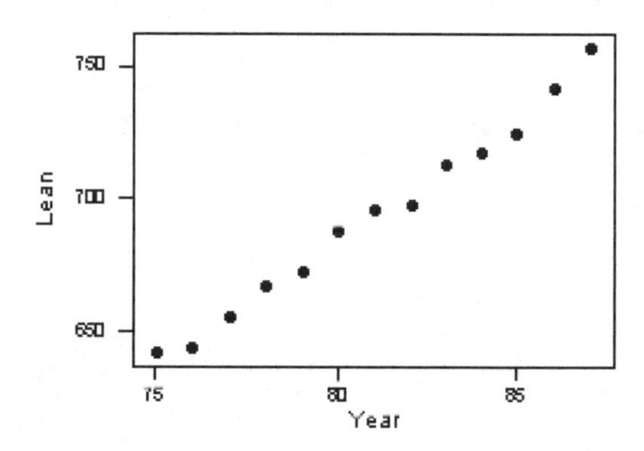

There are no outliers or other unusual points in the plot. The increasing trend in lean appears to be linear over this range of years.

(b) The Minitab output is reproduced below.

Regression Analysis: Lean versus Year

```
The regression equation is
Lean = - 61.1 + 9.32 Year

Predictor        Coef    SE Coef        T       P
Constant       -61.12      25.13    -2.43   0.033
Year           9.3187     0.3099    30.07   0.000

S = 4.181       R-Sq = 98.8%     R-Sq(adj) = 98.7%

Analysis of Variance

Source            DF        SS         MS        F        P
Regression         1     15804      15804   904.12    0.000
Residual Error    11       192         17
Total             12     15997
```

The equation of the least-squares line is given directly as "The regression equation."

```
Lean = - 61.1 + 9.32 Year
```

The values of b_0 and b_1 are given in the output under the column labeled Coef (short for Coefficients) and are given with more significant digits than in the regression equation output.

The percent of variation in lean explained by the line is r^2 and can be computed from the analysis of variance table as

$$r^2 = \text{SSM/SST} = 15{,}804/15{,}997 = 98.8\%$$

The sums of squares SSM and SST can be found in the ANOVA table in the column labeled SS (for sum of squares). SSM is often referred to as the "Regression" sum of squares as well as the model sum of squares, and SST is the "Total" sum of squares. The value of r^2 is also provided directly as R-Sq and is given in the line of output directly above the ANOVA table.

(c) The slope parameter β_1 is the average rate of change of the lean. For a 99% confidence interval we need the upper $(1 - C)/2 = (1 - 0.99)/2 = 0.005$ critical value for the $t(n - 2) = t(13 - 2) = t(11)$ distribution. From Table D we find $t^* = 3.106$. The estimated slope $b_1 = 9.3187$. Its standard error $\text{SE}_{b_1} = 0.3099$ is read from the column labeled SE Coef in the output and the row corresponding to year. The confidence interval is

$$b_1 \pm t^*\text{SE}_{b_1} = 9.3187 \pm (3.106)(0.3099) = 9.3187 \pm 0.9625 = (8.356, 10.281)$$

where the units are in tenths of a millimeter per year.

Exercise 10.40

(a) The predicted value in 1918 ($x^* = 18$) using the equation of the least-squares line is

$$\hat{y} = b_0 + b_1 x^* = -61.1 + (9.32)(18) = 106.66$$

or a lean of 2.9107 meters.

(b) From the plot it is clear that a prediction at $x^* = 18$ is extrapolation. The linear relationship (constant rate of change) between year and lean was found to be reasonable over the time period 1975 through 1987. This prediction assumes that this linear relationship continued to be valid back to 1918, which appears not to be the case.

Exercise 10.41

(a) The year 2009 would be coded as $x = 109$.

(b) The predicted value in 2009 ($x^* = 109$) using the equation of the least-squares line is

$$\hat{y} = b_0 + b_1 x^* = -61.1 + (9.32)(109) = 954.78$$

or a lean of 2.9955 meters.

(c) Since we are interested in the particular value that might occur in 2009 rather than the mean value for that year, we would want to use the margin of error associated with a prediction interval.

Exercise 10.47

```
Analysis of Variance

Source            DF        SS        MS      F       P
Regression         1    3445.9    3445.9   9.50   0.005
Residual Error    28   10152.4     362.6
Total             29   13598.3
```

The degrees of freedom satisfy the relationship DFM + DFE = DFT, or DFE = DFT – DFM = 29 – 1 = 28. Alternatively, we are given the sample size $n = 30$, so we can use the formula DFE = $n - 2 = 30 - 2 =$ 28. The sums of squares satisfy the relationship SSM + SSE = SST, or

$$\text{SSE} = \text{SST} - \text{SSM} = 13{,}598.3 - 3445.9 = 10{,}152.4$$

The mean sum of squares MS is defined by the relation

$$\text{MS} = \frac{\text{sum of squares}}{\text{degrees of freedom}}$$

which gives

$$\text{MSE} = \frac{10{,}152.4}{28} = 362.6$$

Exercise 10.48

Both the regression standard error and r^2 can be computed from the entries in the completed analysis of variance table in Exercise 10.47. The required formulas are contained in Section 10.2 of your text.

$$s = \sqrt{\text{MSE}} = \sqrt{362.6} = 19.04$$

$$r^2 = \frac{\text{SSM}}{\text{SST}} = \frac{3445.9}{13{,}598.3} = 25.3\%$$

Exercise 10.49

Since $\sqrt{\sum (x_i - \bar{x})^2 / n - 1} = 16.45$, we know that

$$\sqrt{\sum (x_i - \bar{x})^2} = 16.45\sqrt{n-1} = 16.45\sqrt{30-1} = 88.59$$

and the value s, the regression standard error, was computed in Exercise 10.47 as s = 19.04. This gives the standard error of the least-squares slope as

$$\text{SE}_{b_1} = \frac{s}{\sqrt{\sum (x_i - \bar{x})^2}} = \frac{19.04}{88.59} = 0.215$$

For a 95% confidence interval we need the upper $(1 - C)/2 = (1 - 0.95)/2 = 0.025$ critical value for the $t(n - 2) = t(30 - 2) = t(28)$ distribution. From Table D we find $t^* = 2.048$. From the Minitab output in the text, the estimated slope is $b_1 = 0.663$ and the confidence interval is

$$b_1 \pm t^* \text{SE}_{b_1} = 0.663 \pm (2.048)(0.215) = 0.663 \pm 0.440 = (0.223,\ 1.103)$$

CHAPTER 11

MULTIPLE REGRESSION

OVERVIEW

Multiple linear regression extends the techniques of simple linear regression to situations involving $p > 1$ explanatory variables $x_1, x_2, \cdots, x_p$. The statistical model for multiple linear regression is

$$y_i = \beta_0 + \beta_1 x_{i1} + \beta_2 x_{i2} + \ldots + \beta_p x_{ip} + \varepsilon_i$$

where $i = 1, 2, \ldots, n$. The deviations ε_i are assumed to be independent and normally distributed with mean 0 and standard deviation σ. The **parameters** of the model are $\beta_0, \beta_1, \beta_2, \cdots, \beta_p$, and σ. The βs are estimated by $b_0, b_1, b_2, \cdots, b_p$ by the **principle of least squares.** The parameter σ is estimated by

$$s = \sqrt{\text{MSE}} = \sum \frac{e_i^2}{n-p-1}$$

where the e_i are the **residuals**

$$e_i = y_i - \hat{y}_i$$

and

$$\hat{y}_i = b_0 + b_1 x_{i1} + b_2 x_{i2} + \ldots + b_p x_{ip}$$

In practice, the b's and s are calculated using statistical software.

A **level *C* confidence interval** for β_j is

$$b_j \pm t^* \text{SE}_{b_j}$$

where t^* is the upper $(1 - C)/2$ critical value for the $t(n - p - 1)$ distribution. SE_{b_j} is the standard error of b_j and in practice is computed using statistical software.

The **test of the hypothesis** H_0: $\beta_j = 0$ is based on the ***t* statistic**

$$t = \frac{b_j}{\text{SE}_{b_j}}$$

with P-values computed from the $t(n - p - 1)$ distribution. In practice statistical software is used to carry out these tests.

In multiple regression, interpretation of these confidence intervals and tests depends on the particular explanatory variables in the multiple regression model. The estimate of β_j represents the effect of the explanatory variable x_j when it is added to a model already containing the other explanatory variables. The test of H_0: $\beta_j = 0$ tells us if the improvement in the ability of our model to predict the response y by adding x_j to a model already containing the other explanatory variables is statistically significant. It does not tell us if x_j would be useful for predicting the response in multiple regression models with a different collection of explanatory variables.

The **ANOVA table** for a multiple regression is analogous to that in simple linear regression. It gives the sum of squares SSM for the model, the sum of squares SSE for error, the total sum of squares SST for all sources of variation, the degrees of freedom DFM, DFE, and DFT for these sums of squares, the mean square MSM for the model, and the mean squares MSE for error. The sums of squares are

$$\text{SSM} = \sum (\hat{y}_i - \bar{y})^2$$

$$\text{SSE} = \sum (y_i - \hat{y}_i)^2$$

$$\text{SST} = \sum (y_i - \bar{y})^2$$

and the degrees of freedom are

$$\text{DFM} = p$$

$$\text{DFE} = n - p - 1$$

$$\text{DFT} = n - 1$$

The mean squares MS are defined by the relation

$$\text{MS} = \frac{\text{sum of squares}}{\text{degrees of freedom}}$$

In practice, these quantities are computed using statistical software. The results are often summarized in an ANOVA table which usually has a form like the following.

Source	Degrees of freedom	Sum of squares	Mean square	F
Model	DFM	SSM	SSM/DFM	MSM/MSE
Error	DFE	SSE	SSE/DFE	
Total	DFT	SST		

The **ANOVA *F*-statistic** is the ratio MSM/MSE and is used to test

$$H_0\text{: } \beta_1 = \beta_2 = \ldots = \beta_p = 0$$

Under H_0, this statistic has an $F(p, n - p - 1)$ distribution, which can be used to compute P-values using Table E. Notice that evidence against H_0 tells us only that at least one of the β_j differs from 0 but not which one. Deciding which $\beta_j \neq 0$ requires further analysis using the procedures for confidence intervals or hypothesis tests for the individual β_j mentioned above, keeping in mind the difficulties of interpreting these individual inferences.

The **squared multiple correlation** can be written as

$$R^2 = \text{SSM/SST}$$

and is interpreted as the proportion of the variability in the response variable y that is explained by the explanatory variables $x_1, x_2, \ldots, x_p$ in the multiple regression.

GUIDED SOLUTIONS

Exercise 11.15

KEY CONCEPTS: Multiple linear regression, R^2, F test, t tests, interpretation of coefficients

(a) Give the null and alternative hypotheses that are tested by each of the t statistics. What are the results of these three significance tests?

(b) For each of the regression coefficients, interpret the sign and give a short explanation of what this means. For example, the negative coefficient of GPA means that higher GPAs are associated with lower marijuana use.

(c) What do the numbers 3 and 85 represent in the expression $F(3, 85)$ and how are they computed?

(d) The null and alternative hypotheses tested by the ANOVA F test are

H_0:

H_a:

What is the value of the F-statistic and the P-value? What do you conclude?

(e) Are there any potential problems using variables measured by having the students complete a questionnaire?

(f) How well can these results be applied to other populations of high school students?

Exercise 11.19

KEY CONCEPTS: Multiple linear regression, *F* test, *t* tests

(a) Use the information in the table to give the equation of the least-squares line.

(b) The null and alternative hypotheses tested by the ANOVA *F* test are

H_0:

H_a:

Interpret the result of this test.

(c) Explain the three entries in the column labeled "*t*."

(d) Give the degrees of freedom associated with each of the *t* statistics. You will need to use the description of the data to first find the values of *n* and *p*.

Exercise 11.41

KEY CONCEPTS: Multiple linear regression, interpretation of the regression coefficients in a multiple linear regression, plots of residuals

(a) Write the model for the analysis including all assumptions.

(b) Summarize the results of your analysis by filling in the ANOVA table following. We recommend that the analysis be carried out using statistical software.

Source	Degrees of freedom	Sum of squares	Mean square	F
Model				
Error				
Total				

The regression equation is

PCB = _____ + _____ PCB52 + _____ PCB118 + _____ PCB138 + _____ PCB180

Variable	Parameter estimate	Std. error	t ratio	P-value
Constant				
PCB52				
PCB118				
PCB138				
PCB180				

$s = \sqrt{\text{MSE}} =$ ________ $R^2 =$ ________

The null and alternative hypotheses tested by the ANOVA F test are

H_0:

H_a:

The P-value of your test of H_0 is ________________ (you may be able to read this directly from the statistical software package you use).

What do you conclude?

Look at the results for the individual coefficients. Can any of the explanatory variables be dropped from the model?

(c) Use statistical software to make a normal quantile plot. Do the data appear to be approximately normal? Are there any outliers? Plot the residuals versus each of the explanatory variables. Are there any patterns?

COMPLETE SOLUTIONS

Exercise 11.15

(a) $H_0: \beta_1 = 0; H_a: \beta_1 \neq 0$. $t = 4.55$ and $P < 0.001$. The regression coefficient for GPA is significantly different from 0.

$H_0: \beta_2 = 0; H_a: \beta_2 \neq 0$. $t = 2.69$ and $P < 0.01$. The regression coefficient for Popularity is significantly different from 0.

$H_0: \beta_3 = 0; H_a: \beta_3 \neq 0$. $t = 2.69$ and $P < 0.01$. The regression coefficient for Depression is significantly different from 0.

(b) Higher GPAs are associated with lower marijuana usage while higher popularity scores and higher depression scores are each associated with greater marijuana usage.

(c) The numbers 3 and 85 are the numerator and denominator degrees of freedom, respectively, in the F distribution. They are calculated as the number of explanatory variables $p = 3$ for the numerator degrees of freedom and $n - p - 1 = 89 - 3 - 1 = 85$ for the denominator degrees of freedom.

(d) The null and alternative hypotheses tested by the ANOVA F test are

$$H_0\text{: } \beta_1 = \beta_2 = \beta_3 - 0 \text{ and } H_a\text{: at least one of } \beta_1, \beta_2, \text{ and } \beta_3 \text{ is not } 0$$

We are told that $F = 14.83$ $P < 0.001$. There is strong evidence that the regression coefficients for GPA, Popularity, and Depression are not all 0.

(e) In this type of study, there is always a danger that students will lie or make errors in their responses.

(f) The student population in the suburban Florida high school where the data was collected is quite different from an inner city student population in Washington, D.C. or a rural student population in Wyoming. The ranges of the predictors and response will be different in different high schools (more or less marijuana use, higher or lower GPAs, etc.) and the relationships among the variables may be different as well.

Exercise 11.19

(a) Overall = 3.33 + 0.82 × Unfavorable + 0.57 × Favorable

(b) The null and alternative hypotheses tested by the ANOVA F test are

$$H_0\text{: } \beta_1 = \beta_2 = 0$$

$$H_a\text{: at least one of } \beta_1 \text{ and } \beta_2 \text{ is not } 0$$

where β_1 is the regression coefficient of "Unfavorable" and β_2 is the regression coefficient of "Favorable" in our multiple linear regression model.

The value of the F-statistic is 33.7. There are $n = 152$ observations and $p = 2$ explanatory variables. Under H_0, this statistic has an $F(p, n - p - 1) = F(2, 149)$ distribution. We are told that the P-value is less than 0.01. Using Table E, we find that the P-value is actually less than 0.001. We conclude that there is strong evidence that at least one of the two regression coefficients is different from 0 in the population regression equation.

(c) These are the t statistics for testing the hypotheses H_0: $\beta_j = 0$ for $j = 0$, 1, and 2. When $j = 0$, the t for "Constant" is testing if the intercept is 0, while for $j = 1$ and 2, the t's are testing that the slopes for "Unfavorable" and "Favorable" are 0. The t's for each of the slopes are large, indicating that both explanatory variables contribute to the model, given the other variable is in the model.

(d) The degrees of freedom associated with each of the t statistics are $n - p - 1 = 152 - 2 - 1 = 149$.

Exercise 11.41

(a) The statistical model for multiple linear regression is

$$y_i = \beta_0 + \beta_1 x_{i1} + \beta_2 x_{i2} + \beta_3 x_{i3} + \beta_4 x_{i4} + \varepsilon_i$$

where $i = 1, 2, \ldots, n$. The deviations ε_i are assumed to be independent and normally distributed with mean 0 and standard deviation σ.

(b) A summary of the results of a multiple linear regression analysis are given below. The numbers were obtained from Minitab but have been reformatted.

Source	Degrees of freedom	Sum of squares	Mean square	F	P-value
Model	4	237,246	59,311	1456.18	0.000
Error	64	2,607	41		
Total	68	239,853			

The regression equation is

PCB = 0.94 + 11.9 PCB52 + 3.76 PCB118 + 3.88 PCB138 + 4.18 PCB180

Variable	Parameter estimate	Std. error	t ratio	P-value
Constant	0.937	1.229	0.76	0.449
PCB52	11.8727	0.7290	16.29	0.000
PCB118	3.7611	0.6424	5.85	0.000
PCB138	3.8842	0.4978	7.80	0.000
PCB180	4.1823	0.4318	9.69	0.000

$s = \sqrt{\text{MSE}} = 6.382$ $\quad R^2 = 98.9\%$

The null and alternative hypotheses tested by the ANOVA F test are

$$H_0: \beta_1 = \beta_2 = \beta_3 = \beta_4 = 0$$

$$H_a: \text{at least one of } \beta_1, \beta_2, \beta_3, \text{or } \beta_4 \text{ is not } 0$$

The P-value of your test of H_0 is 0.000, which provides very strong support for the alternative hypothesis that at least one of β_1, β_2, β_3, or β_4 is not 0. Looking at the t ratios and P-values for the individual coefficients suggests that none of the four explanatory variables can be dropped from the model.

However, the explanatory variables are highly correlated with each other. A model with only PCB52 and PCB138 as the explanatory variables has an R^2 of 97.3%, which is almost as large as the R^2 for the model containing all four variables. See Exercise 11.43 of the text for more information on alternative models.

(c) The normal probability plot of the residuals is given below. It shows two outliers, which are circled, but there appear to be no other major departures from the assumption of normality.

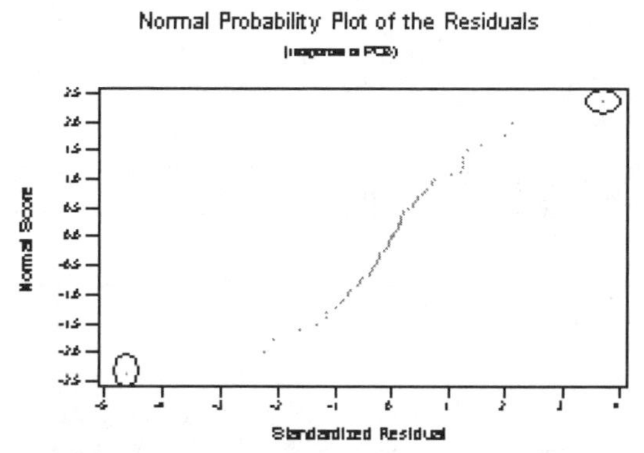

The plots of the residuals against the four explanatory variables are given below. The outliers are apparent in the plots, but otherwise there are no obvious patterns that would suggest a problem with the model.

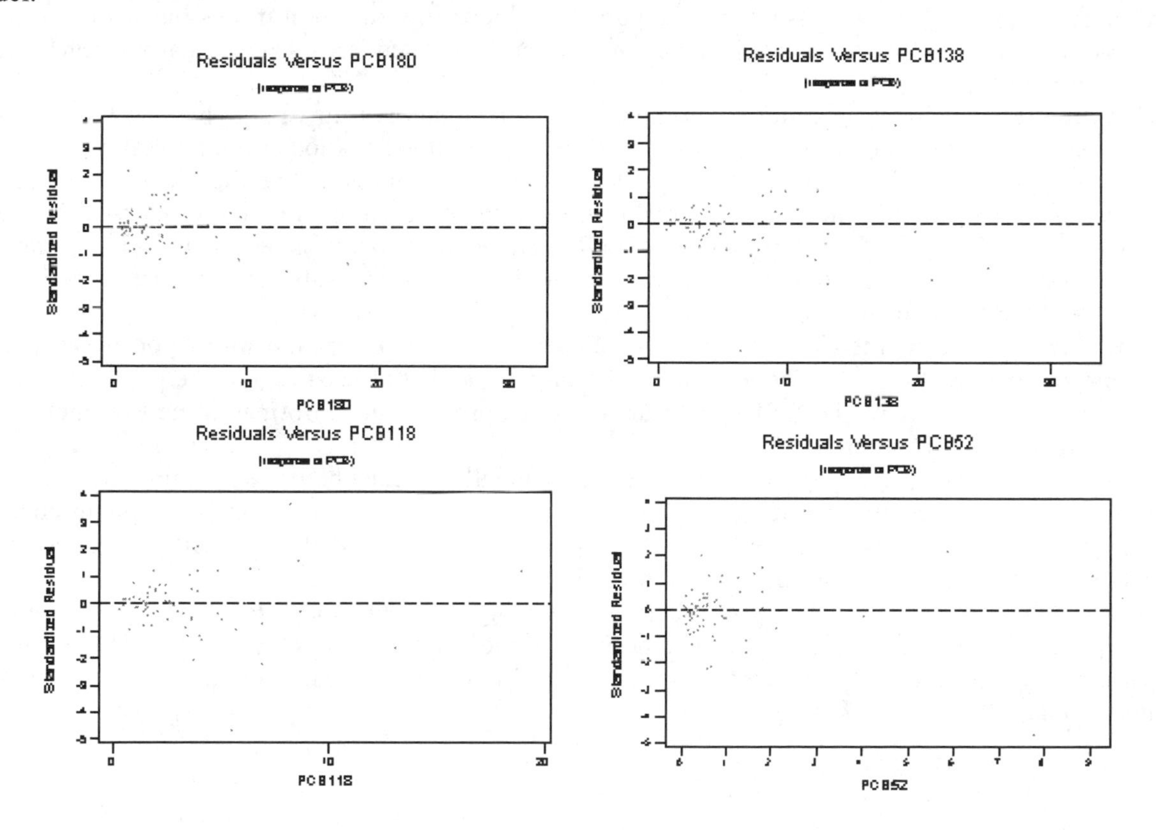

CHAPTER 12

ONE-WAY ANALYSIS OF VARIANCE

OVERVIEW

One-way analysis of variance is a generalization of the two-sample t procedures. It allows comparison of more than two populations based on independent SRSs from each. As in the pooled two-sample t procedures, the populations are assumed to be normal with possibly different means but with a common standard deviation. In one-way analysis of variance we are interested in making formal inferences about the population means.

The simplest graphical procedure to compare the populations is to give side-by-side boxplots (see Chapter 1). Normal quantile plots can be used to check for extreme deviations from normality or outliers. The summary statistics required for the analysis of variance calculations are the means and standard deviations of each of the samples. An informal procedure to check the assumption of equal variances is to make sure that the ratio of the largest to the smallest standard deviation is less than 2. If the standard deviations satisfy this criterion and the normal quantile plots seem satisfactory, then the one-way ANOVA is an appropriate analysis.

The ***F*-statistic** computed in the ANOVA table can be used to test the **null hypothesis** that the population means are all equal. The **alternative hypothesis** is that at least two of the population means are not equal. Rejection of the null hypothesis does not provide any information as to which of the population means are different.

If the researcher has specific questions about the population means before examining the data, these questions can often be expressed in terms of **contrasts.** Tests and confidence intervals about contrasts provide answers to these questions and allow the researcher to say more about which population means are different and what the sizes of these differences are.

When there are no specific questions before examining the data, **multiple comparisons** are often used to follow up rejection of the null hypothesis in a one-way analysis of variance. These multiple comparisons are designed to determine which pairs of population means are different and to give confidence intervals for the differences.

GUIDED SOLUTIONS

Exercise 12.39

KEY CONCEPTS: Assumptions in one-way ANOVA, one-way ANOVA calculations, multiple comparisons

(a) Use software to complete the table below. Fill in the sample sizes, means, and standard deviations for the three treatments. This provides numerical summary statistics for the data.

Variable	Treatment	N	Mean	StDev
BMD	Control			
	Low dose			
	High dose			

The simplest graphical method to describe this data set would be side-by-side boxplots. Use software to draw side-by-side boxplots.

(b) Examine the assumptions necessary for ANOVA. Use the rule for examining standard deviations in ANOVA to determine if it is reasonable to use a pooled standard deviation for the analysis of these data. If your software makes normal quantile plots, then draw one for each group. Does the assumption of normality seem satisfied?

(c) Use your software to run the ANOVA. Fill in the table below and summarize your findings.

One-way ANOVA: BMD versus Treatment

Source	DF	SS	MS	F	P-value
Treatment					
Error					
Total					

(d) Multiple comparison methods are designed to determine which pairs of means are different. These methods are used when the null hypothesis is rejected, and we didn't have specific questions about the means in advance of the analysis. Multiple comparison procedures are performed by computing t statistics for all pairs of means using the formula

$$t_{ij} = \frac{\overline{x}_i - \overline{x}_j}{s_p\sqrt{\frac{1}{n_i} + \frac{1}{n_j}}}$$

This is the same t statistic that would be used when studying the contrast $\psi = \mu_i - \mu_j$. We declare the population means μ_i and μ_j different whenever

$$|\, t_{ij} \,| \geq t^{**}$$

where the value of t^{**} depends on the multiple-comparisons procedure being applied. In this exercise, we will use the Bonferroni multiple-comparisons procedure and the value of $t^{**} = 2.49$. When doing the calculations, it is often simplest to summarize them in a table like the one below. For the three groups we let C = Control, L = Low dose, and H = High dose. The details for the first pair of means (C, L) are worked out for you. Note that the quantity $s_p\sqrt{\frac{1}{n_i} + \frac{1}{n_j}}$ which appears in the calculations of the t statistic has the same value for any two groups since the sample sizes are all the same. Use of this fact will save some time in the calculations.

For the pair (C, L),

$$\overline{x}_{\text{L}} - \overline{x}_{\text{C}} = 0.21887 - 0.21593 = 0.00294$$

$$s_p\sqrt{\frac{1}{n_{\text{L}}} + \frac{1}{n_{\text{C}}}} = 0.01437\sqrt{\frac{1}{15} + \frac{1}{15}} = 0.00525$$

and

$$t_{\text{LC}} = \frac{\overline{x}_{\text{L}} - \overline{x}_{\text{C}}}{s_p\sqrt{\frac{1}{n_{\text{L}}} + \frac{1}{n_{\text{C}}}}} = \frac{0.00294}{0.00525} = 0.56$$

Complete the table below and interpret the results.

Pair of means	$\overline{x}_i - \overline{x}_j$	$s_p\sqrt{\frac{1}{n_i} + \frac{1}{n_j}}$	t_{ij}
(L, C)	0.00294	0.00525	0.56
(H, C)			
(H, L)			

(e) Using your results in parts (c) and (d), write a short report explaining the effect of kudzu isoflavones on the femur of the rat.

Exercise 12.47

KEY CONCEPTS: Assumptions in one-way ANOVA, one-way ANOVA calculations, F-statistic

(a) Use software to complete the table below. Fill in the sample sizes, means, and standard deviations for the three groups.

Variable	Group	N	Mean	StDev
Density	Control			
	Highjump			
	Lowjump			

Use the rule for examining standard deviations in ANOVA to determine if it is reasonable to use a pooled standard deviation for the analysis of these data.

(b) Use your software to run the ANOVA and then fill in the table below.

One-way ANOVA: Density versus Group

Source	DF	SS	MS	F	P
Group					
Error					
Total					

Give the F-statistic, its degrees of freedom, and P-value. What do you conclude?

Exercise 12.54

KEY CONCEPTS: Summary statistics, one-way ANOVA calculations, *F*-statistic

(a) Use software to complete the table below. Fill in the sample sizes, means, standard deviations, and standard errors for the subjects in each of the four groups.

Variable	Group	N	Mean	SE Mean	StDev
Flaking	Keto				
	Placebo				
	PyrI				
	PyrII				

Make a graph of the means on the axes below.

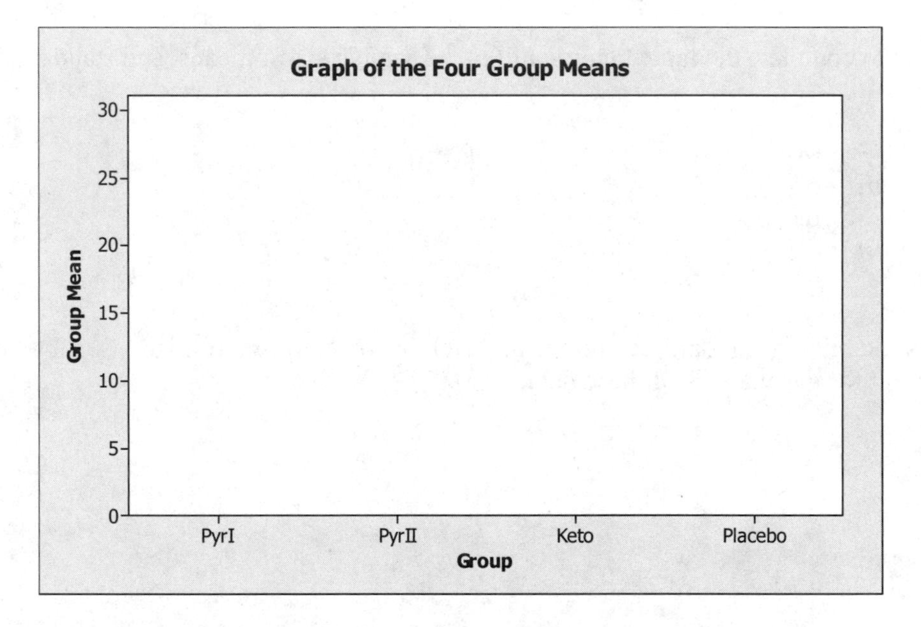

(b) Use your software to run the ANOVA and then fill in the table below.

One-way ANOVA: Flaking versus Group

Source	DF	SS	MS	F	P
Group					
Error					
Total					

Write a short summary of the results and your conclusions.

Exercise 12.57

KEY CONCEPTS: Comparisons among the means in one-way ANOVA, contrasts

(a) Contrasts are comparisons among the means designed to answer specific questions. These questions are posed before the data are collected. There are four means in this problem: μ_{Plac}, μ_{PyrI}, μ_{PyrII}, and μ_{Keto}. Give each of the contrasts as a combination of some of these means. Remember that the sum of the coefficients of the means will add to zero for a contrast. The first contrast is completed below, and the remaining two are left for you to do on your own.

(1) Placebo versus the average of the three treatments

The contrast is $\psi_1 = \mu_{\text{Plac}} - \frac{1}{3}(\mu_{\text{PyrI}} + \mu_{\text{PyrII}} + \mu_{\text{Keto}})$. The four coefficients are 1, –1/3, –1/3, and –1/3, which add to zero.

(2) Keto versus the average of the two Pyr treatments

$\psi_2 =$

(3) PyrI versus PyrII

$\psi_3 =$

(b) To help you do this problem we reproduce the data for the four groups below.

Variable	Group	N	Mean	SE Mean	StDev
Flaking	Keto	106	16.028	0.0904	0.931
	Placebo	28	29.393	0.301	1.595
	PyrI	112	17.393	0.108	1.142
	PyrII	109	17.202	0.130	1.352

The estimate of a contrast or combination of population means $\psi = \sum a_i \mu_i$ is given by the same combination of the sample means. The sample contrast or estimator of ψ is $c = \sum a_i \bar{x}_i$. The standard error of the contrast is given by

$$\text{SE}_c = s_p \sqrt{\sum \frac{a_i^2}{n_i}}$$

The first thing that needs to be done to apply these results in a particular setting is to identify the a_i in the contrast. The details for the first contrast and its standard error are given on the next page and the remaining two are left for you to do on your own.

(1) Placebo versus the average of the three treatments

The contrast is $\psi_1 = \mu_{\text{Plac}} - \frac{1}{3}(\mu_{\text{PyrI}} + \mu_{\text{PyrII}} + \mu_{\text{Keto}})$. The estimate of this contrast is

$$c = \sum a_i \bar{x}_i = \bar{x}_{\text{Plac}} - \frac{1}{3}(\bar{x}_{\text{PyrI}} + \bar{x}_{\text{PyrII}} + \bar{x}_{\text{Keto}}) = 29.393 - \frac{1}{3}(17.393 + 17.202 + 16.028) = 12.519$$

The standard error of a contrast requires the within-groups estimate of the standard deviation s_p and the a_i and n_i for the contrast. The within-groups estimate of the standard deviation s_p corresponds to the square root of the MSE in the ANOVA table. The ANOVA table for these data is given in Exercise 12.54, and $s_p = \sqrt{\text{MSE}} = 1.196$. The a_i for this contrast are 1, –1/3, –1/3, and –1/3. These a_i correspond to sample sizes n_i of 28, 112, 109, and 106, respectively.

We can now apply the general formula for the standard error of a contrast.

$$\text{SE}_c = s_p\sqrt{\sum \frac{a_i^2}{n_i}} = 1.196\sqrt{\sum \frac{(1)^2}{28} + \frac{(-1/3)^2}{112} + \frac{(-1/3)^2}{109} + \frac{(-1/3)^2}{106}} = 0.236$$

(2) Keto versus the average of the two Pyr treatments

$$c = \sum a_i \bar{x}_i =$$

$$\text{SE}_c = s_p\sqrt{\sum \frac{a_i^2}{n_i}} =$$

(3) PyrI versus PyrII

$$c = \sum a_i \bar{x}_i =$$

$$\text{SE}_c = s_p\sqrt{\sum \frac{a_i^2}{n_i}} =$$

(c) A test of the null hypothesis that the contrast is 0 uses the t statistic given by $t = \frac{c}{\text{SE}_c}$. The t has 351 degrees of freedom and critical values can be approximated with the z table. Using your results from part (b), evaluate the t statistics for each of the three contrasts and then summarize the results of your tests and your conclusions. The details for the first significance test are given below, and the remaining two are left for you to do on your own.

(1) Placebo versus the average of the three treatments

From part (b), the contrast is $c = 12.519$ and $\text{SE}_c = 0.236$. The t statistic for this contrast is

$$t = \frac{c}{\text{SE}_c} = \frac{12.519}{0.236} = 53.05.$$

The P-value is essentially 0, showing that the placebo is associated with higher mean flaking than the average of the three treatments. The remaining contrasts explore the differences among the treatments.

(2) Keto versus the average of the two Pyr treatments

$$t = \frac{c}{\text{SE}_c} =$$

(3) PyrI versus PyrII

$$t = \frac{c}{\text{SE}_c} =$$

Summarize the results of your tests and your conclusions below.

COMPLETE SOLUTIONS

Exercise 12.39

(a) The table below provides the sample sizes, means, and standard deviations for the three treatments.

Variable	Treatment	N	Mean	StDev
BMD	Control	15	0.21887	0.01159
	Low dose	15	0.21593	0.01151
	High dose	15	0.23507	0.01877

The side-by-side boxplots for the three treatments are given below.

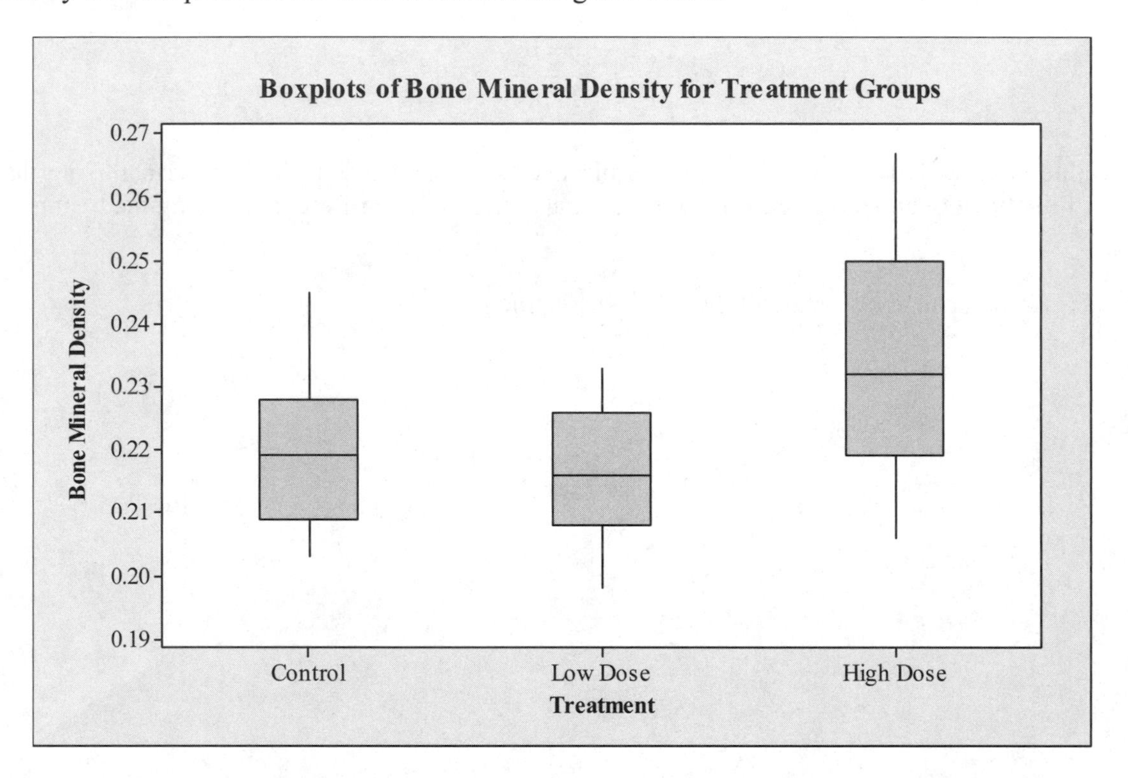

The summary statistics and plots both show the low-dose group and control to be similar. Both have similar means and standard deviations, and the boxplots for the two groups show similar medians and spreads. The high-dose group has a higher mean (and median) and greater spread, as shown by the standard deviation as well as the boxplot.

(b) The rule for examining standard deviations in ANOVA is to take the ratio of the largest standard deviation and the smallest standard deviation. If this ratio is less than 2, we can use the methods based on the assumption of equal standard deviations and the results will be approximately correct. In this problem the ratio is $0.01877/0.01151 = 1.63$, so we can assume the standard deviations are approximately equal. The normal quantile plots for the three groups are given on the next page. The assumption of normality seems reasonable for all three groups, and the boxplots in part (a) show no outliers in any of the groups. Additionally, the ANOVA F test is relatively insensitive to moderate nonnormality and unequal variances, especially when the sample sizes are similar. In this case the three sample sizes are equal.

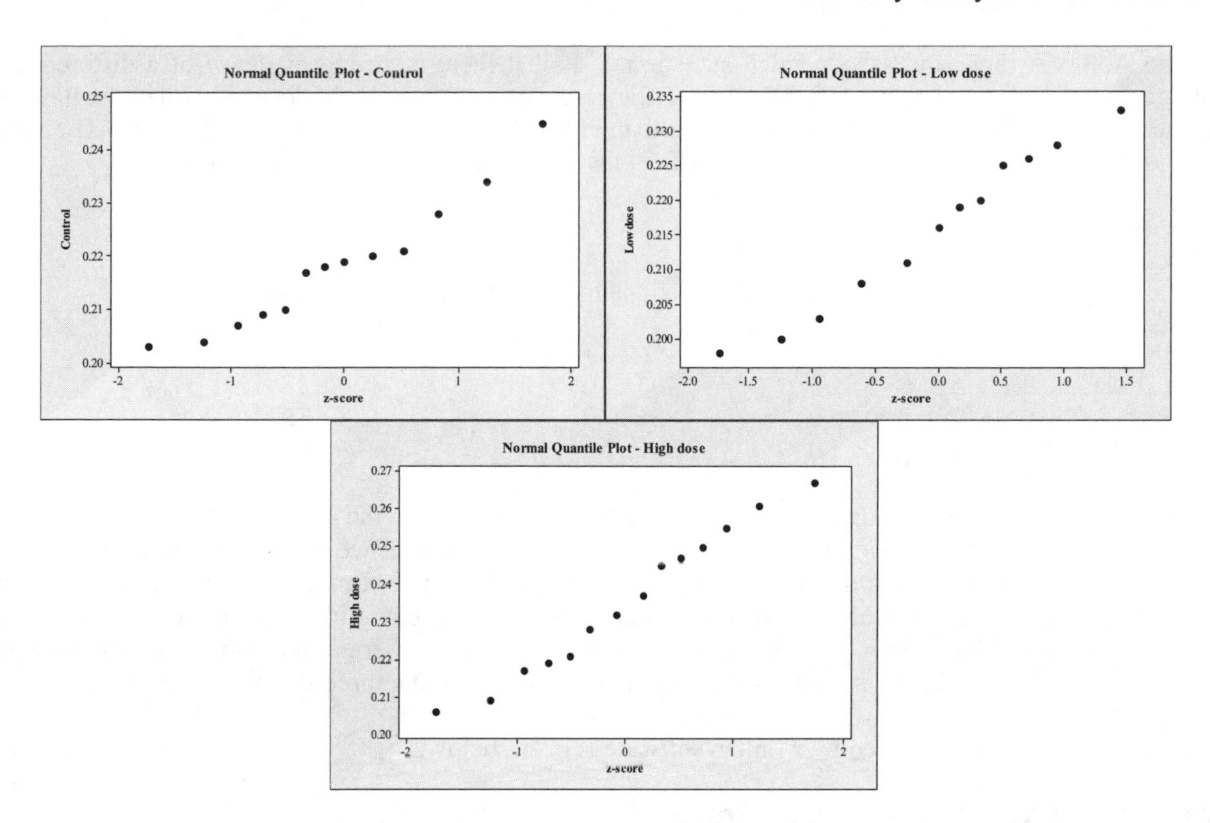

(c) The ANOVA table is given below. The F-statistic is 7.72 with a P-value of 0.001. There is very strong evidence of a difference in the means among the three groups.

One-way ANOVA: BMD versus Treatment

Source	DF	SS	MS	F	P
Treatment	2	0.003186	0.001593	7.72	0.001
Error	42	0.008668	0.000206		
Total	44	0.011853			

(d) The calculations for the three multiple comparisons are summarized in the table below.

Pair of means	$\bar{x}_i - \bar{x}_j$	$s_p\sqrt{\dfrac{1}{n_i}+\dfrac{1}{n_j}}$	t_{ij}
(L, C)	0.00294	0.00525	0.56
(H, C)	0.0162	0.00525	3.08
(H, L)	0.0191	0.00525	3.63

With the critical value of $t^{**} = 2.49$, we find that the high-dose group shows evidence of a higher mean than the control group ($t = 3.08$) and a higher mean than the low-dose group ($t = 3.63$). The means of the control and low-dose groups show no evidence of a difference ($t = 0.56$). These results are quite apparent from the boxplots provided in part (a).

(e) The ANOVA table and associated F-statistic tell us that there is strong evidence of a difference in average bone mineral density (BMD) among the three groups. A high dose of kudzu isoflavones produces a higher BMD in the femur of the rat than either the low-dose or the control group. The mean BMD of the low-dose group and the mean BMD of the control group show little difference.

Exercise 12.47

(a) The sample sizes, means, and standard deviations for the three groups are given below.

Variable	Group	N	Mean	StDev
Density	Control	10	601.10	27.36
	High jump	10	638.70	16.59
	Low jump	10	612.50	19.33

The rule for examining standard deviations in ANOVA is to take the ratio of the largest standard deviation and the smallest standard deviation. If this ratio is less than 2, we can use the methods based on the assumption of equal standard deviations and the results will be approximately correct. In this problem the ratio is 27.36/16.59 = 1.65, so we can assume the standard deviations are approximately equal. Additionally, the ANOVA F test is relatively insensitive to moderate nonnormality and unequal variances, especially when the sample sizes are similar. In this case the three sample sizes are equal.

(b) The ANOVA table produced by Minitab software is given below.

One-way ANOVA: Density versus Group

Source	DF	SS	MS	F	P
Group	2	7434	3717	7.98	0.002
Error	27	12580	466		
Total	29	20013			

The F-statistic is 7.98 with 2 degrees of freedom in the numerator and 27 degrees of freedom in the denominator. The P-value is 0.002, so we reject H_0 and conclude that the population means are not all the same.

Exercise 12.54

.

(a) The sample sizes, means, standard deviations, and standard errors for the subjects in each of the four groups are given below.

Variable	Group	N	Mean	SE Mean	StDev
Flaking	Keto	106	16.028	0.0904	0.931
	Placebo	28	29.393	0.301	1.595
	PyrI	112	17.393	0.108	1.142
	PyrII	109	17.202	0.130	1.352

The graph of the means is given on the next page.

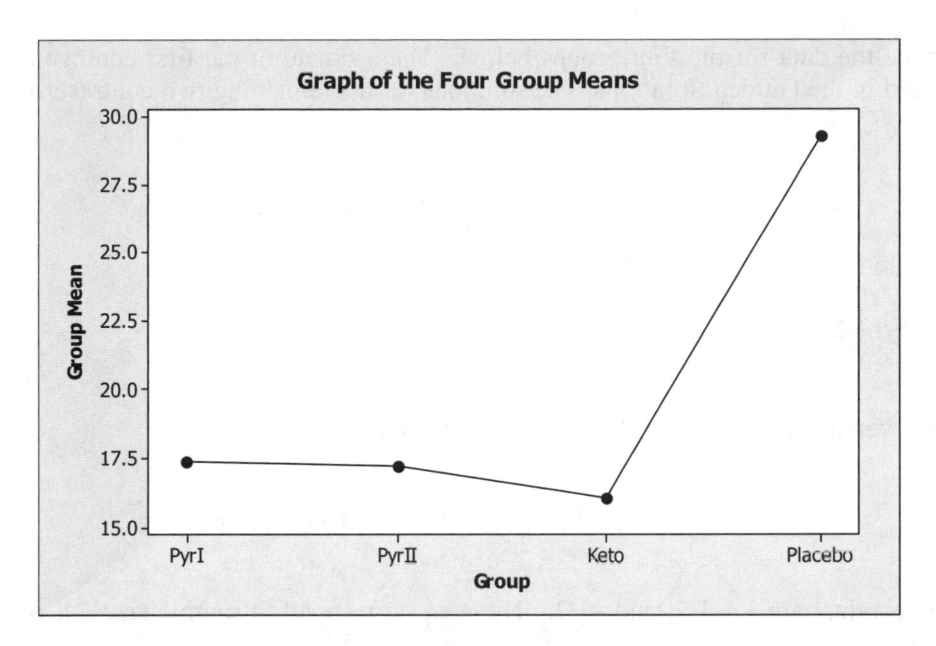

(b) The ANOVA table produced by Minitab software is given below.

One-way ANOVA: Flaking versus Group

Source	DF	SS	MS	F	P
Group	3	4151.43	1383.81	967.82	0.000
Error	351	501.87	1.43		
Total	354	4653.30			

The F-statistic is used to test the null hypothesis that the mean flaking for the four groups is the same, and the alternative is that there are some differences among the four means. For these data, $F = 967.82$ with 3 degrees of freedom in the numerator and 351 degrees of freedom in the denominator. The P-value is approximately 0, so we reject H_0 and conclude that the population means are not all the same. From the graph of the means it appears that the three treatment groups all have less flaking than the placebo. The differences among the three treatments appear to be smaller. To complete the analysis of these data requires either multiple comparisons or analysis of a suitable set of contrasts. Exercise 12.57 from the text suggests several natural contrasts and the solution to this exercise appears below.

Exercise 12.57

(a) The first contrast is given in the Guided Solutions preceding.

(2) Keto versus the average of the two Pyr treatments

$$\psi_2 = \mu_{\text{Keto}} - \frac{1}{2}(\mu_{\text{PyrI}} + \mu_{\text{PyrII}})$$

(3) PyrI versus PyrII

$$\psi_3 = \mu_{\text{PyrI}} - \mu_{\text{PyrII}}$$

(b) We reproduce the data for the four groups below. The estimate of the first contrast and its standard error are provided in the Guided Solutions. The solutions for the remaining two contrasts are given below. Recall that $s_p = \sqrt{\text{MSE}} = 1.196$.

Variable	Group	N	Mean	SE Mean	StDev
Flaking	Keto	106	16.028	0.0904	0.931
	Placebo	28	29.393	0.301	1.595
	PyrI	112	17.393	0.108	1.142
	PyrII	109	17.202	0.130	1.352

(2) Keto versus the average of the two Pyr treatments

$$c = \sum a_i \bar{x}_i = \bar{x}_{\text{Keto}} - \frac{1}{2}(\bar{x}_{\text{PyrI}} + \bar{x}_{\text{PyrII}}) = 16.028 - \frac{1}{2}(17.393 + 17.202) = -1.269$$

The a_i for this contrast are 1, –1/2, and –1/2. These a_i correspond to sample sizes n_i of 106, 112, and 109, respectively.

$$\text{SE}_c = s_p \sqrt{\sum \frac{a_i^2}{n_i}} = 1.196\sqrt{\sum \frac{(1)^2}{106} + \frac{(-1/2)^2}{112} + \frac{(-1/2)^2}{109}} = 0.141$$

(3) PyrI versus PyrII

$$c = \sum a_i \bar{x}_i = \bar{x}_{\text{PyrI}} - \bar{x}_{\text{PyrII}} = 17.393 - 17.202 = 0.191$$

The a_i for this contrast are 1 and –1. These a_i correspond to sample sizes n_i of 112 and 109, respectively.

$$\text{SE}_c = s_p \sqrt{\sum \frac{a_i^2}{n_i}} = 1.196\sqrt{\sum \frac{(1)^2}{112} + \frac{(-1)^2}{109}} = 0.161$$

(c) The significance test for the first contrast was done in the Guided Solutions. The remaining two significance tests are given below

(2) Keto versus the average of the two Pyr treatments

$$t = \frac{c}{\text{SE}_c} = \frac{-1.269}{0.141} = 9$$

(3) PyrI versus PyrII

$$t = \frac{c}{\text{SE}_c} = \frac{0.191}{0.161} = 1.18$$

The t statistic for testing no difference between the means of the flaking measure for PyrI and PyrII has a value of $t = 1.18$ with 351 degrees of freedom. The value of 351 for the degrees of freedom can be found in the ANOVA table and corresponds to the degrees of freedom for error. Using the z table as an approximation to the t with 351 degrees of freedom, we have a P-value of $2 \times 0.119 = 0.238$, since the test is two-sided. This indicates that there is no evidence of a difference in the flaking measure when shampooing with pyrithione zinc shampoo once or twice. The t statistic for testing no difference between the mean flaking measure of Keto and the average of the means of the flaking measure for PyrI and PyrII has a value of $t = 9$ with 351 degrees of freedom. This gives a P-value that is essentially 0, showing that the ketoconazole shampoo reduces the flaking measure compared with the average of the means of the pyrithione zinc shampoo used once and twice. Finally, the t statistic for testing no difference between the placebo and the average of the means of the three treatments has a value of $t = 53.27$ with 351 degrees of freedom. This gives a P-value that is essentially 0, showing that the average of the means of the flaking measure for the three treatments is less than the placebo. Confidence intervals for these contrasts would provide information about the size of the reduction in the flaking measure associated with these contrasts.

CHAPTER 13

TWO-WAY ANALYSIS OF VARIANCE

OVERVIEW

Two-way analysis of variance is designed to compare the means of populations that are classified according to two factors. As with one-way ANOVA, the populations are assumed to be normal with possibly different means and the same standard deviation. The observations are independent SRSs drawn from each population.

The preliminary data examination should include examination of means and standard deviations and normal quantile plots. Typically the means are summarized in a two-way table with the rows corresponding to the levels of one factor and the columns corresponding to the levels of the second factor. The **marginal means** are computed by taking averages of these cell means across the rows and columns. These means are typically plotted so that the **main effects** of each factor can be examined as well as their **interaction.**

In the two-way ANOVA table, the **model** variation is broken down into parts due to each of the main effects and a third part due to the interaction. In addition, the ANOVA table organizes the calculations required to compute *F*-statistics and *P*-values to test hypotheses about these main effects and their interaction. The within-group variance is estimated by pooling the standard deviations from the cells and corresponds to the mean square for error in the ANOVA table.

GUIDED SOLUTIONS

Exercise 13.21

KEY CONCEPTS: Plotting group means, main effects, interactions

(a) Complete the plot of the mean GITH for these diets on the axes provided on the next page. You should plot the two means when Chromium is Low above the letter L, and the two means when Chromium is Normal above the letter N. Then, for each Eat group, connect the points for the two Chromium means. Label the two resulting lines with an M and an R, for the two Eat groups.

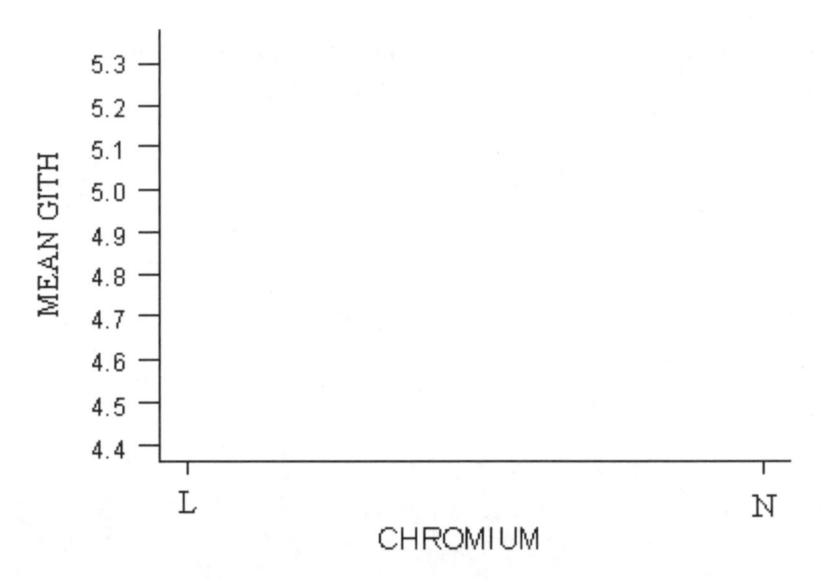

(b) Try to identify some of the main features of the plot. Are the lines approximately parallel? Is the effect of Chromium similar for both levels of Eat? Are the lines for the two levels of Eat far apart? Is there much change in the mean GITH when going from Low to Normal levels of Chromium? Think about what the answers to these different questions are telling you about the factors Eat and Chromium.

(c) In the table below fill in the marginal means.

	Eat	
Chromium	M	R
L	4.545	5.175
N	4.425	5.317

Compare the difference between the M and R diets for each level of Chromium. How does this comparison show up in your plot?

Exercise 13.25

KEY CONCEPTS: Computing group means, interaction plots, two-way ANOVA

(a) In the table below fill in the sample sizes, means, and standard deviations for the 12 material-time groups. You should try and do the computations using computer software.

Group	N	Mean	StDev
ECM1,4			
ECM1,8			
ECM2,4			
ECM2,8			
ECM3,4			
ECM3,8			
MAT1,4			
MAT1,8			
MAT2,4			
MAT2,8			
MAT3,4			
MAT3,8			

Is it reasonable to pool the variances?

(b) Plot the means of the 12 groups computed in (a) on the axes given. The time is on the *x* axis and the mean % Gpi is on the *y* axis. For each material, connect the two means corresponding to the different times.

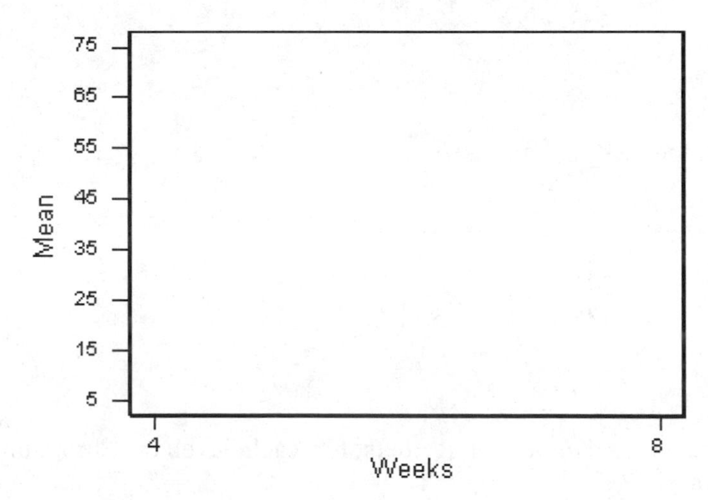

Describe any patterns you see. Focus on the differences between times, materials, and any interaction between them.

(c) Use your computer software to generate an ANOVA table similar to the one on the next page.

```
Analysis of Variance for % Gpi

Source           DF        SS        MS        F        P
Material          5   27045.1    5409.0   251.26    0.000
Time              1       6.2       6.2     0.29    0.595
Material*Time     5       6.2       1.3     0.06    0.998
Error            24     516.7      21.5
Total            35   27574.3
```

Are the results of the significance tests in agreement with the impressions that you formed from the graphs in (b)? If the results are not what you expected, try to come up with a reason for the discrepancies between the ANOVA table and the plots.

Exercise 13.31

KEY CONCEPTS: Computing group means, interaction plots, two-way ANOVA

(a) In the table below fill in the sample sizes, means, and standard deviations for the 15 tool-time groups. You should try and do the computations using computer software.

Tool	Time	N	Mean	Stdev
1	1			
1	2			
1	3			
2	1			
2	2			
2	3			
3	1			
3	2			
3	3			
4	1			
4	2			
4	3			
5	1			
5	2			
5	3			

(b) The plot on the next page is for the means of each of the time and tool combinations. The tool is on the x axis and the mean diameter is on the y axis. For each of the three times, connect the five means corresponding to the different tools. This type of plot is most easily drawn using software. What is suggested by the plot? Does there appear to be an interaction? Are there differences among the tools or an effect of time?

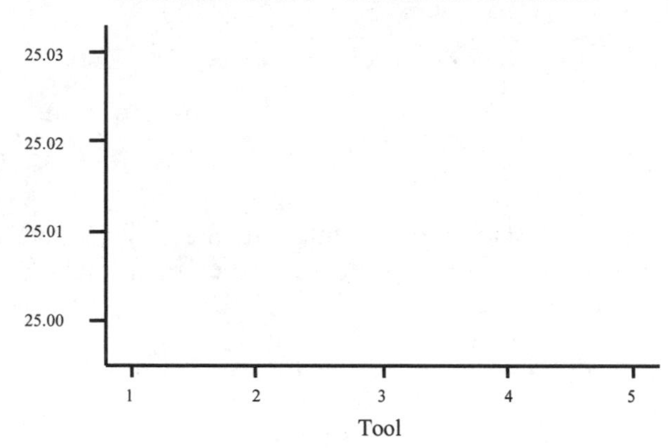

(c) Use your computer software to generate an ANOVA table similar to the one below.

Analysis of Variance for Diameter

Source	DF	SS	MS	F	P
Tool	4	0.00359714	0.00089928	412.98	0.000
Time	2	0.00018992	0.00009496	43.61	0.000
Tool*Time	8	0.00013324	0.00001665	7.65	0.000
Error	30	0.00006533	0.00000218		
Total	44	0.00398562			

Report the test statistics, degrees of freedom, and P-values for the tests for interaction and main effects.

(d) Write a short paragraph summarizing your results. Be sure to use both the ANOVA in part (c) and the plots in part (b).

COMPLETE SOLUTIONS

Exercise 13.21

(a)

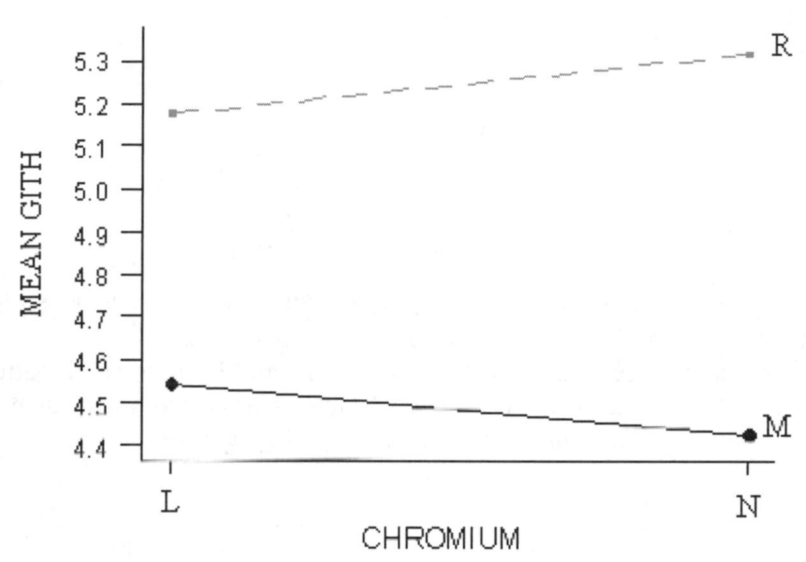

(b) The plot suggests that there may be an interaction. The effect of Chromium when going from Low to Normal levels is to decrease the mean GITH when the rats could eat as much as they wanted (M) and to increase the mean GITH when the total amount the rats could eat was restricted (R). Without a formal hypothesis test, we don't know if this apparent interaction is due to chance variation or whether the effect is real. In terms of the effect of Chromium, it appears to be small compared to the effect of Eat. The two lines are quite far apart which shows that Eat has a large effect. The change in going from the Low to Normal levels of Chromium is much smaller.

(c)

	Eat		Mean
Chromium	M	R	
L	4.545	5.175	4.860
N	4.425	5.317	4.871
Mean	4.485	5.246	4.866

At the Low level of Chromium the difference between M and R is –0.63, and at the Normal level of Chromium the difference between M and R is –0.892. In the plot the two means at the Low level of Chromium are closer together than the two means at the Normal level of Chromium. This is reflected in the fact that the two lines are not parallel.

Exercise 13.25

(a) The table below gives the sample sizes, means, and standard deviations for the 12 material-time groups.

Variable	Group	N	Mean	StDev
% Gpi	ECM1,4	3	65.00	8.66
	ECM1,8	3	63.33	2.89
	ECM2,4	3	63.33	2.89
	ECM2,8	3	63.33	5.77
	ECM3,4	3	73.33	2.89
	ECM3,8	3	73.33	5.77
	MAT1,4	3	23.33	2.89
	MAT1,8	3	21.67	5.77
	MAT2,4	3	6.67	2.89
	MAT2,8	3	6.67	2.89
	MAT3,4	3	11.67	2.89
	MAT3,8	3	10.00	5.00

The ratio of the largest to smallest standard deviation is $8.66/2.89 = 3.00 > 2$, which is a larger ratio of the standard deviations than we would like to see when we pool the variances.

The table below summarizes the means for the different material-by-time combinations and provides the marginal means for material and time. This type of table is produced easily using most computer software packages and makes it easier to draw the interaction plot in part (b).

ROWS: TIME COLUMNS: MATERIAL

	ECM1	ECM2	ECM3	MAT1	MAT2	MAT3	ALL
4	65.000	63.333	73.333	23.333	6.667	11.667	40.556
8	63.333	63.333	73.333	21.667	6.667	10.000	39.722
ALL	64.167	63.333	73.333	22.500	6.667	10.833	40.139

(b)

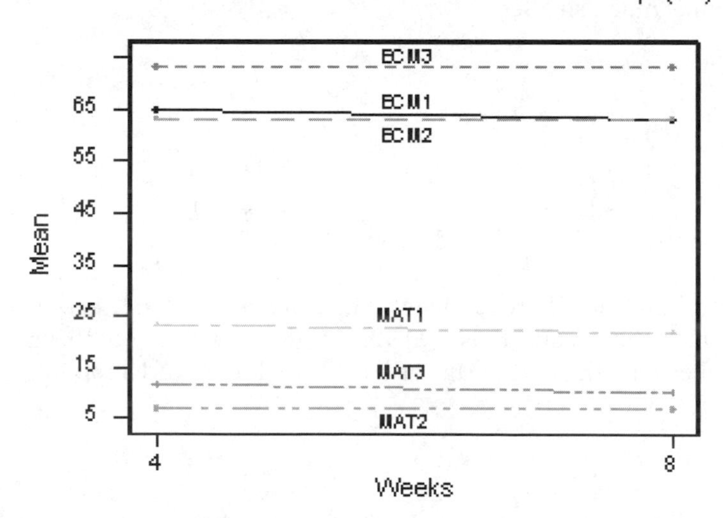

The most striking feature of the plot is the complete lack of a time effect. The % Gpi means at 4 and 8 weeks are almost identical, and as a result there is almost no interaction. The important effect is the material, with the ECM (extracellular) materials having a much higher % Gpi than the MAT (inert) materials.

(c) The ANOVA table is reproduced below.

Analysis of Variance for % Gpi

Source	DF	SS	MS	F	P
Material	5	27045.1	5409.0	251.26	0.000
Time	1	6.2	6.2	0.29	0.595
Material*Time	5	6.2	1.3	0.06	0.998
Error	24	516.7	21.5		
Total	35	27574.3			

The ANOVA table supports the interaction plot in (b). There is no evidence of a time effect or an interaction between time and material. However, there is a highly significant effect of material. Note that the ANOVA table does not tell which materials are different or whether there is a difference between the ECM materials and the MAT materials. Further multiple comparisons or examination of specific contrasts as in one-way ANOVA are required to determine these potential differences.

Exercise 13.31

(a) The table below gives the means and standard deviations for each of the 15 treatment combinations.

Tool	Time	N	Mean	Stdev
1	1	3	25.031	0.001
1	2	3	25.028	0.000
1	3	3	25.026	0.000
2	1	3	25.017	0.002
2	2	3	25.020	0.001
2	3	3	25.016	0.000
3	1	3	25.006	0.002
3	2	3	25.013	0.001
3	3	3	25.009	0.001
4	1	3	25.012	0.000
4	2	3	25.019	0.001
4	3	3	25.014	0.004
5	1	3	24.997	0.001
5	2	3	25.006	0.000
5	3	3	25.000	0.002

(b) The plot below is for the means of each of the time and tool combinations.

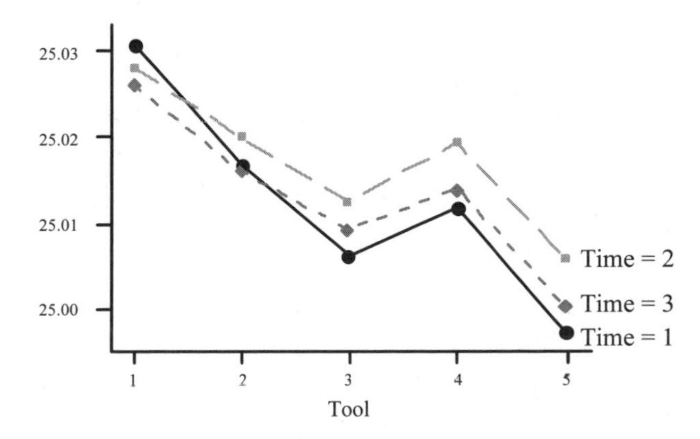

The plot suggests little interaction, as the lines for each time appear fairly parallel. The biggest difference is between the tools, but this is of little importance because we expect the tools to have slightly different diameters and this can be adjusted. There is also a difference in times. With the exception of tool 1, time 2 has the largest diameters, time 3 has the middle diameters, and time 1 has the smallest diameters.

(c) The analysis of variance table below was produced by Minitab. Both main effects and the interaction are statistically significant. The degrees of freedom for the main effect of tool are 4 and 30, for the main effect of time are 2 and 30, and for interaction are 8 and 30. The *F*-statistics and *P*-values are also given in the table.

Analysis of Variance for Diameter

Source	DF	SS	MS	F	P
Tool	4	0.00359714	0.00089928	412.98	0.000
Time	2	0.00018992	0.00009496	43.61	0.000
Tool*Time	8	0.00013324	0.00001665	7.65	0.000
Error	30	0.00006533	0.00000218		
Total	44	0.00398562			

(d) Both main effects and the interaction are statistically significant, although the *F* for interaction is much smaller than the *F* for the main effect of time, which is much smaller than the *F* for the main effect of tool. The interaction occurs because tool 1's mean diameters changed differently over time compared to the other tools. The difference in the sizes of these effects was apparent in the plot in part (b). The reason that the interaction is statistically significant despite the appearance of the plot is that there is so little variability within each treatment. Many of the standard deviations were zero, as shown in part (a). When there is very little within-group variability or the sample sizes are large, it is possible to detect effects that, while being statistically significant, are of little practical importance. Because tool differences are of limited interest, the most important finding is the fact that the diameters vary with time.

CHAPTER 14

LOGISTIC REGRESSION

OVERVIEW

In Chapters 5 and 8 we studied random variables y which can take only two values (yes or no, success or failure, live or die, acceptable or not). It is usually convenient to code the two values as 0 (failure) and 1 (success) and let p denote the probability of a 1 (success). If y is observed on n independent trials, the total number of 1s can often be modeled as binomial with n trials and probability of success p. In this chapter, we consider methods that allow us to investigate how y depends on one or more explanatory variables. The methods of simple linear and multiple linear regression do not directly apply since the distribution of y is binomial rather than normal. However, it is possible to use a method, called **logistic regression,** which is similar to simple and multiple linear regression.

We define the **odds** as $p/(1-p)$, the ratio of the probability that the event happens to the probability that the event does not happen. If $\hat{p}$ is the sample proportion, the sample odds are $\hat{p}/(1-\hat{p})$. The **logistic regression model** relates the (natural) log of the odds to the explanatory variables. In the case of a single explanatory variable x, the logistic regression model is

$$\log\left(\frac{p}{1-p}\right) = \beta_0 + \beta_1 x$$

where the responses y_i, for $i = 1, \ldots, n$, are n independent binomial random variables with parameters 1 and p_i; that is, they are independent with distributions $B(1, p_i)$. The **parameters** of this logistic regression model are β_0 (the intercept of the logistic model) and β_1 (the slope of the logistic model). The quantity e^{β_1} is called the **odds ratio.**

The formulas for estimates of the parameters of the logistic regression model are, in general, very complicated, and so, in practice, the estimates are computed using statistical software. Such software typically gives estimates b_0 for β_0 and b_1 for β_1 along with estimates SE_{b_0} and SE_{b_1} for the standard errors of these estimates. A **level C confidence interval for the intercept** β_0 is then determined by the formula

$$b_0 \pm z^* SE_{b_0}$$

where z^* is the upper $(1 - C)/2$ quantile of the standard normal distribution. Similarly, a **level C confidence interval for the slope** β_1 is determined by the formula

$$b_1 \pm z^* \, \mathrm{SE}_{b_1}$$

A **level C confidence interval for the odds ratio** e^{β_1} is obtained by transforming the confidence interval for the slope, yielding the formula

$$\left(e^{b_1 - z^* \mathrm{SE}_{b_1}}, e^{b_1 + z^* \mathrm{SE}_{b_1}}\right)$$

To **test the hypotheses** H_0: $\beta_1 = 0$ and H_a: $\beta_1 \neq 0$ compute the **test statistic**

$$X^2 = \left(\frac{b_1}{\mathrm{SE}_{b_1}}\right)^2$$

Since the random variable X^2 has approximately a χ^2 distribution with 1 degree of freedom, the P-value for this test is $P(\chi^2 \geq X^2)$. This is the same as testing the null hypothesis that the odds ratio is 1.

In **multiple logistic regression,** the response variable again has two possible values, but there can be more than one explanatory variable. Multiple logistic regression is analogous to multiple linear regression. Fitting multiple logistic regression models is done, in practice, with statistical software.

GUIDED SOLUTIONS

Exercise 14.25

KEY CONCEPTS: Odds, odds ratio

(a) You need to compute

$$\hat{p}_{\text{high}} = \frac{\text{number in high blood pressure group who died of cardiovascular disease}}{\text{total number of men in the high blood pressure group}}$$

$$=$$

and then

$$\text{ODDS} = \frac{\hat{p}_{\text{high}}}{1 - \hat{p}_{\text{high}}} =$$

(b) Repeat the type calculations you did in (a), but now for the men with low blood pressure.

$$\hat{p}_{\text{low}} =$$

$$\text{ODDS} =$$

(c) In this setting, recall that the odds ratio is

$$\text{Odds ratio} = e^{\beta_1} = \frac{\text{Odds for men with high blood pressure}}{\text{Odds for men with low blood pressure}} =$$

Interpret this odds ratio in words.

Exercise 14.27

KEY CONCEPTS: Logistic regression, confidence interval for the slope, test that the slope is 0

(a) A 95% confidence interval for the slope β_1 is determined by the formula $b_1 \pm z^* \text{SE}_{b_1}$, where z^* is the upper 0.025 quantile of the standard normal distribution. Look up z^* in Table A and then, using the information given in the problem, compute

$$b_1 \pm z^* \text{SE}_{b_1} =$$

(b) Recall that the X^2 statistic for testing the null hypothesis that the slope is zero is $X^2 = \left(\frac{b_1}{\text{SE}_{b_1}}\right)^2$ and has a χ^2 distribution with 1 degree of freedom. Use the values for b_1 and SE_{b_1} given in the problem to compute this quantity.

$$X^2 = \left(\frac{b_1}{\text{SE}_{b_1}}\right)^2 =$$

Now use Table F to find the approximate *P*-value:

$$P\text{-value} = P(\chi^2 \geq X^2) =$$

(c) Now summarize your results and conclusions. What can you conclude about the probability of death from cardiovascular disease for men with high blood pressure versus that for men with low blood pressure?

Exercise 14.29

KEY CONCEPTS: Confidence interval for the odds ratio

(a) Recall that a level *C* confidence interval for the odds ratio e^{β_1} is obtained by transforming the level *C* confidence interval for the slope, $b_1 \pm z^* \text{SE}_{b_1}$, to yield the formula

$$\left(e^{b_1 - z^*\text{SE}_{b_1}}, e^{b_1 + z^*\text{SE}_{b_1}}\right)$$

Referring to Exercise 14.27(a), what is the 95% confidence interval for the slope? Hence, what is a 95% confidence interval for the odds ratio?

(b) What does the interval you found in (a) tell you about the odds that a man with high blood pressure dies from cardiovascular disease relative to the odds that a man with low blood pressure dies from cardiovascular disease?

Exercise 14.40

KEY CONCEPTS: Multiple logistic regression

(a) You will need access to statistical software that allows you to do multiple logistic regression. SAS, SPSS, Minitab (later versions), and JMP will allow you to do multiple logistic regression. Consult your software's manual for details and explanation of the output. The output should include the value of X^2 (or z) for testing that the coefficients of both the explanatory variables are 0 and will probably also give the *P*-value of the test. If the *P*-value is not given, refer to Table F in the text. The appropriate degrees of freedom in this case is 2.

(b) Your statistical software should provide estimates of each of the coefficients and the standard errors of these coefficients. 95% confidence intervals can then be constructed using the formula

$$\text{parameter estimate} \pm z^*(\text{standard error of parameter estimate})$$

(c) What do you conclude based on your findings in (a) and (b)? Would you use both SATM and SATV to predict HIGPA, just one, or neither?

Complete Solutions

Exercise 14.25

(a) The proportion of men who died from cardiovascular disease in the high blood pressure group is

$$\hat{p}_{\text{high}} = \frac{\text{number in high blood pressure group who died of cardiovascular disease}}{\text{total number of men in the high blood pressure group}} = \frac{55}{3338} = 0.0165$$

The odds are

$$\text{ODDS} = \frac{\hat{p}_{\text{high}}}{1-\hat{p}_{\text{high}}} = \frac{55/3338}{1-(55/3338)} = \frac{55/3338}{3283/3338} = \frac{55}{3283} = 0.0168$$

(b) The proportion of men who died from cardiovascular disease in the low blood pressure group is

$$\hat{p}_{\text{low}} = \frac{\text{number in low blood pressure group who died of cardiovascular disease}}{\text{total number of men in the low blood pressure group}} = \frac{21}{2676} = 0.0078$$

The odds are

$$\text{ODDS} = \frac{\hat{p}_{\text{low}}}{1-\hat{p}_{\text{low}}} = \frac{21/2676}{1-(21/2676)} = \frac{21/2676}{2655/2676} = \frac{21}{2655} = 0.0079$$

(c) The odds ratio, with the odds for the high blood pressure group in the numerator, is

$$\text{Odds ratio} = \frac{\text{ODDS for men with high blood pressure}}{\text{ODDS for men with low blood pressure}} = \frac{55/3283}{21/2655} = 2.1181$$

This tells us that the odds that, in the study group, a man with high blood pressure dies from cardiovascular disease are slightly more than twice the odds that a man with low blood pressure dies from cardiovascular disease.

Exercise 14.27

(a) We are given in the problem that $b_1 = 0.7505$ and $\text{SE}_{b_1} = 0.2578$. For a 95% confidence interval, from Table A we have $z^* = 1.96$, and so

$$b_1 \pm z^* \text{SE}_{b_1} = 0.7505 \pm 1.96(0.2578) = 0.7505 \pm 0.5052$$

or equivalently, the interval (0.2453, 1.2557).

(b) We compute

$$X^2 = \left(\frac{b_1}{\text{SE}_{b_1}}\right)^2 = \left(\frac{0.7505}{0.2578}\right)^2 = (2.9112)^2 = 8.4751$$

Using Table F for the χ^2 distribution with 1 degree of freedom, we estimate the P-value to be

$$P\text{-value} = P(\chi^2 \geq X^2) = P(\chi^2 \geq 8.4751)$$

which is between 0.0025 and 0.005.

(c) There is strong evidence that the slope parameter in the logistic regression model is different from 0 (or, equivalently, that the odds ratio is different from 1). In addition, we are 95% confident that the slope parameter has a value between 0.2453 and 1.2557. Although the problem does not say so, these results, along with those of Exercise 14.25, suggest that the explanatory variable must be an indicator variable with value 1 for men with high blood pressure and value 0 for men with low blood pressure. Since a positive slope parameter implies that the probability of death from cardiovascular disease increases as the explanatory variable increases, we may conclude that the data provide strong evidence that the probability of death from cardiovascular disease is higher for men with high blood pressure than for men with low blood pressure.

Exercise 14.29

(a) We saw in (a) of Exercise 14.27 that a 95% confidence interval for the slope in the logistic regression model is (0.2453, 1.2557). Hence, a 95% confidence interval for the odds ratio is

$$\left(e^{b_1 - z^* \mathrm{SE}_{b_1}}, e^{b_1 + z^* \mathrm{SE}_{b_1}}\right) = \left(e^{0.2453}, e^{1.2557}\right) = (1.2780, 3.5103)$$

(b) We are 95% confident that the odds that a man with high blood pressure dies from cardiovascular disease is between 1.2780 and 3.5103 times higher than the odds that a man with low blood pressure dies from cardiovascular disease.

Exercise 14.40

(a) The results of running a multiple logistic regression model to predict HIGPA from SATM and SATV are given below. We used JMP to perform the analysis.

Response: HIGPA
Converged by Gradient
Whole-Model Test

Model	–LogLikelihood	DF	ChiSquare	Prob>ChiSq
Difference	7.11016	2	14.22031	0.0008
Full	140.55964			
Reduced	147.66980			

The information regarding the hypothesis test that the coefficients for both explanatory variables are zero is found in the row labeled "Difference" in the table above. The entry under the column labeled "ChiSquare" gives the value of the test statistic X^2. This test statistics has value $X^2 = 14.22031$ and has 2 degrees of freedom (as indicated in the column labeled "DF"). The *P*-value of the test is given in the column labeled "Prob>ChiSq" and is 0.0008. We would reject the null hypothesis that the coefficients for both explanatory variables are zero and conclude that at least one of the explanatory variables is helpful for predicting the odds that HIGPA is 1 or, equivalently, that GPA $\geq$ 3.00.

(b) The information for the individual parameter estimates is summarized below.

Response: HIGPA
Converged by Gradient
Parameter Estimates

Term	Estimate	Std Error	ChiSquare	Prob>ChiSq
Intercept	4.54290625	1.1617665	15.29	<.0001
SATM	–0.00369	0.0019136	3.72	0.0538
SATV	–0.003527	0.0017518	4.05	0.0441

The estimates of the coefficients for SATM and SATV are given in the table above in the column labeled "Estimate." The standard errors of these estimates are given in the column labeled "Std Error." Recall that the general formula for a 95% confidence interval is

parameter estimate $\pm$ z^*(standard error of parameter estimate)

with $z^* = 1.96$ for a 95% confidence interval. Using the information in the table we find the following:

95% CI for coefficient of SATM:

$-0.00369 \pm 1.96(0.0019136) = -0.00369 \pm 0.00375 = (-0.00744, +0.00006)$

95% CI for coefficient of SATV:

$-0.003527 \pm 1.96(0.0017518) = -0.003527 \pm 0.003434 = (-0.006961, -0.000093)$

(c) For predicting the odds that HIGPA is 1 or, equivalently, the odds that GPA $\geq$ 3.00, the results of (b) show that the 95% confidence interval for the coefficient of SATV does not contain 0; hence, the coefficient would be declared significantly different from 0 at the 0.05 level when SATV is added to a model containing SATM. This suggests that SATV improves prediction when added to a model already containing SATM. However, the 95% confidence interval for the coefficient of SATM just barely contains 0, and we would not declare the coefficient to be significantly different from 0 at the 0.05 level when added to a model containing SATV. Since 0 is very near the upper limit of the 95% confidence interval for SATM, we might nevertheless be inclined to include SATM in the model. This is further supported by the fact that the *P*-value for the test that both coefficients are 0 is quite small. The data suggest that using both SATM and SATV as predictors is reasonable.

CHAPTER 15

NONPARAMETRIC TESTS

SECTION 15.1

OVERVIEW

Many of the statistical procedures described in previous chapters assumed that the samples were drawn from normal populations. **Nonparametric tests** do not require any specific form for the distributions of the populations from which the samples were drawn. Many nonparametric tests are **rank tests;** that is, they are based on the **ranks** of the observations rather than on the observations themselves. When ranking the observations from smallest to largest, tied observations receive the average of their ranks.

The **Wilcoxon rank sum test** compares two distributions. The objective is to determine if one distribution has systematically larger values than the other. The observations are ranked, and the **Wilcoxon rank sum statistic** W is the sum of the ranks of one of the samples. The Wilcoxon rank sum test can be used in place of the **two-sample** t **test** when samples are small or the populations are far from normal.

Exact P-values require special tables and are produced by some statistical software. However, many statistical software packages give only approximate P-values based on a normal approximation, typically with a continuity correction employed. Many packages also make an adjustment in the normal approximation when there are ties in the ranks.

GUIDED SOLUTION

Exercise 15.9

KEY CONCEPTS: Ranking data, two-sample problem, Wilcoxon rank sum test

(a) Order the observations from smallest to largest in the space provided. Use a different color or underline those observations in the high-progress group. This will make it easier to determine the ranks assigned to each group.

(b) Now suppose the first sample is the high-progress group and the second sample is the low-progress group. The choice of which sample we call the first sample and which we call the second sample is arbitrary. However, the Wilcoxon rank sum statistic W is the sum of the ranks of the first sample, and the formulas for the mean and variance of W distinguish between the sample sizes for the first and the second samples. What are the ranks of the high-progress observations? Use these ranks to compute the value of W.

$W =$

What are the values of n_1, n_2, and N? Use these to evaluate the mean and standard deviation of W according to the formulas below.

$$\mu_W = \frac{n_1(N+1)}{2} =$$

$$\sigma_W = \sqrt{\frac{n_1 n_2 (N+1)}{12}} =$$

(c) What kind of values would W have if the alternative were true? Use the normal approximation (with the continuity correction) to find the approximate P-value. If you have access to software or tables to evaluate the exact P-value, compare it with the approximation.

$$z = \frac{W - \mu_W}{\sigma_W} =$$

P-value =

What are your conclusions?

(d) Order the observations from smallest to largest in the space provided. Use a different color or underline those observations in the high-progress group. This will make it easier to determine the ranks assigned to each group. Remember that ties are broken by assigning all tied values the average of the ranks they occupy.

COMPLETE SOLUTION

Exercise 15.9

(a) The observations are first ordered from smallest to largest. The observations written in bold are from the high-progress group.

0.28, 0.38, 0.49, **0.54,** 0.66, 0.77, **0.79, 0.80, 0.82, 0.89**

(b) Suppose the first sample is the high-progress group and the second sample is the low-progress group. In this case, $n_1 = n_2 = 5$, and N, the sum of the sample sizes, is 10. The high-progress observations in bold have ranks 4, 7, 8, 9, and 10, and the sum of these ranks is $W = 38$. The values for the mean and variance are

$$\mu_W = \frac{n_1(N+1)}{2} = \frac{5(10+1)}{2} = 27.5$$

$$\sigma_W = \sqrt{\frac{n_1 n_2(N+1)}{12}} = \sqrt{\frac{(5)(5)(10+1)}{12}} = 4.787$$

(c) We expect W to be large when the alternative hypothesis is true, as the high-progress group should receive the larger ranks. If we use the continuity correction, we act as if the value of $W = 38$ occupies the interval from 37.5 to 38.5. This means that, to compute the P-value, we first calculate the probability that $W \geq 37.5$, giving

$$P(W \geq 37.5) = P\left(\frac{W - \mu_W}{\sigma_W} \geq \frac{37.5 - 27.5}{4.787}\right) = P(Z \geq 2.09) = 0.0183$$

The exact value using statistical software is 0.016, so the agreement is quite good when using the normal approximation with the continuity correction. There is strong evidence that the high-progress readers tend to have higher scores than the low-progress readers.

(d) The observations are first ordered from smallest to largest. The observations written in bold are from the high-progress group.

0.00, 0.36, 0.40, 0.55, **0.55, 0.57, 0.70, 0.72,** 0.72, **0.84**

The ranks assigned to these observations are

1, 2, 3, 4.5, **4.5, 6, 7, 8.5,** 8.5, **10**

where ties have been broken using average ranks.

SECTION 15.2

OVERVIEW

The **Wilcoxon signed rank test** is a nonparametric test for matched pairs. It tests the null hypothesis that there is no systematic difference between the observations within a pair against the alternative that one observation tends to be larger.

The test is based on the **Wilcoxon signed rank statistic W^+,** which provides another example of a nonparametric test using ranks. The absolute values of the observations are ranked and the sum of the ranks of the positive (or negative) differences gives the value of W^+. The **matched pairs *t* test** and the **sign test** are two other alternative tests for this setting.

P-values can be found from special tables of the distribution or a normal approximation to the distribution of W^+. Some software computes the exact P-value and other software uses the normal approximation, typically with a ties correction. Many packages make an adjustment in the normal approximation when there are ties in the ranks.

GUIDED SOLUTION

Exercise 15.26

KEY CONCEPTS: Matched pairs, Wilcoxon signed rank statistic

First, give the null and alternative hypotheses:

H_0:
H_a:

To compute the Wilcoxon signed rank statistic, first order the absolute values of the differences and rank them. Due to the large number of ties in this exercise, you need to be careful when computing the ranks. For any tied group of observations, they should each receive the average rank for the group. (Note that the positive observations are in bold.) In addition, you need to drop the observations that are equal to zero (the pretest and the posttest scores were the same) and reduce the sample size accordingly. What is the new value of n?

Absolute values	Ranks
1	1.5
1	1.5
2	4.0
2	4.0
2	4.0
3	8.5
3	8.5
3	8.5
3	8.5
3	8.5
3	8.5
6	14.5
6	14.5
6	14.5
6	14.5
6	14.5
6	14.5

To see how the ranks are computed, the 1s would get ranks 1 and 2, so their average rank is 1.5. The 2s would get ranks 3, 4, and 5, so their average rank is 4.0, and so on. If W^+ is the sum of the ranks of the positive observations, compute the value of W^+.

$W^+ =$

Evaluate the mean and standard deviation of W^+ according to the formulas below.

$$\mu_{W^+} = \frac{n(n+1)}{4} =$$

$$\sigma_{W^+} = \sqrt{\frac{n(n+1)(2n+1)}{24}} =$$

Now use the mean and standard deviation to compute the standardized rank sum statistic:

$$z = \frac{W^+ - \mu_{W^+}}{\sigma_{W^+}} =$$

Do you expect W^+ to be small or large if the alternative is true? Use the normal approximation to find the approximate *P*-value. (Note: The continuity correction is often used for statistics whose values are whole numbers. This is true for W^+ when there are no tied observations. However, when there are tied observations and W^+ can take on values that are not whole numbers, the continuity correction would not be employed.)

What are your conclusions?

COMPLETE SOLUTION

Exercise 15.26

The null and alternative hypotheses are

H_0: scores have the same distribution for the pretest and posttest
H_a: posttest scores are systematically higher than pretest scores

The Wilcoxon signed rank statistic is

$$W^+ = 1.5 + 1.5 + 4.0 + 4.0 + 4.0 + 8.5 + 8.5 + 8.5 + 8.5 + 8.5 + 8.5 + 14.5 + 14.5 + 14.5 + 14.5 + 14.5 = 138.5$$

The values for the mean and variance are (recall that three observations had a value of zero and were dropped, reducing n to 17)

$$\mu_{W^+} = \frac{n(n+1)}{4} = \frac{17(17+1)}{4} = 76.5$$

$$\sigma_{W^+} = \sqrt{\frac{n(n+1)(2n+1)}{24}} = \sqrt{\frac{(17)(18)(34+1)}{24}} = 21.125$$

and the standardized signed rank statistic W^+ is

$$z = \frac{W^+ - \mu_{W^+}}{\sigma_{W^+}} = \frac{138.5 - 76.5}{21.125} = 2.93$$

If posttest scores are systematically higher, we would expect the differences (posttest – pretest) to be positive. Thus, the ranks of the positive observations should be large and we would expect the value of the statistic W^+ to be large when the alternative hypothesis is true. The approximate *P*-value is $P(Z \geq 2.93) = 0.0017$.

The output from the Minitab computer package gives a similar result. Many computer packages, including Minitab, include a correction to the standard deviation in the normal approximation to account for the ties in the ranks.

```
Wilcoxon Signed Rank Test

TEST OF MEDIAN = 0.000000 VERSUS MEDIAN G.T. 0.000000

                    N FOR   WILCOXON
             N      TEST    STATISTIC    P-VALUE
CHANGE      17       17       138.5       0.002
```

The data show that the scores on understanding of spoken French have improved after attending a summer language institute.

SECTION 15.3

OVERVIEW

The **Kruskal-Wallis test** is the nonparametric test for the **one-way analysis of variance** setting. In comparing several populations, it tests the null hypothesis that the distribution of the response variable is the same in all groups and the alternative hypothesis that some groups have distributions of the response variable that are systematically larger than others.

The **Kruskal-Wallis statistic** H compares the average ranks received for the different samples. If the alternative is true, some of these should be larger than others. Computationally, it essentially arises from applying the usual one-way ANOVA to the ranks of the observations rather than to the observations themselves.

P-values can be found from special tables of the distribution or a chi-square approximation to the distribution of H. When the sample sizes are not too small, the distribution of H for comparing I populations has approximately a chi-square distribution with $I - 1$ degrees of freedom. Some software computes the exact *P*-value and other software uses the chi-square approximation, typically with an adjustment in the chi-square approximation when there are ties in the ranks.

GUIDED SOLUTION

Exercise 15.34

KEY CONCEPTS: One-way ANOVA, Kruskal-Wallis statistic

(a) The Kruskal-Wallis test is testing

H_0: the distribution of insects trapped is the same for all colors
H_a: some colors have systematically higher numbers of trapped insects

What are the values of I, the n_i and N in this problem?

(b) To compute the Kruskal-Wallis test statistic, the 20 observations are first arranged in increasing order. That step has been carried out below, where we have kept track of the group for each observation. You need to fill in the ranks in the line provided. Remember to use average ranks for tied observations.

```
Trapped     7      11      12      13      14      14      15
Group    Blue    Blue   White   White   White    Blue   Green
Rank


Trapped    16      17      17      20      21      21      25
Group    Blue   White   White    Blue    Blue   White   Green
Rank


Trapped    32      37      38      39      41      45      46
Group   Green   Green   Lemon   Green   Green   Lemon   Lemon
Rank

Trapped    47      48      59
Group   Lemon   Lemon   Lemon
Rank
```

Now fill in the table below to give the ranks for each of the colors and the sum of ranks for each group.

Color	Ranks	Sum of ranks
Lemon		
White		
Green		
Blue		

Use the sum of ranks for the four groups and the numerical values of the n_i and N obtained in part (a) to calculate the Kruskal-Wallis statistic below.

$$H = \frac{12}{N(N+1)} \sum \frac{R_i^2}{n_i} - 3(N+1) =$$

The value of H is compared with critical values in Table F for a chi-square distribution with $I - 1$ degrees of freedom, where I is the number of groups. What is the P-value and what do you conclude?

COMPLETE SOLUTION

Exercise 15.34

(a) $I = 4$, the n_i are each 6, and $N = 24$.

(b) The computations required for the Kruskal-Wallis test statistic are summarized in the tables below.

```
Trapped     7      11     12     13     14     14     15
Group    Blue    Blue  White  White  White   Blue  Green
Rank        1       2      3      4    5.5    5.5      7


Trapped    16      17     17     20     21     21     25
Group    Blue   White  White   Blue   Blue  White  Green
Rank        8     9.5    9.5     11   12.5   12.5     14


Trapped    32      37     38     39     41     45     46
Group   Green   Green  Lemon  Green  Green  Lemon  Lemon
Rank       15      16     17     18     19     20     21

Trapped    47      48     59
Group   Lemon   Lemon  Lemon
Rank       22      23     24
```

Color	Ranks	Sum of ranks
Lemon	17, 20, 21, 22, 23, 24	127
White	3, 4, 5.5, 9.5, 9.5, 12.5	44
Green	7, 14, 15, 16, 18, 19	89
Blue	1, 2, 5.5, 8, 11, 12.5	40

$$H = \frac{12}{N(N+1)}\sum\frac{R_i^2}{n_i} - 3(N+1)$$

$$= \frac{12}{24(24+1)}\left(\frac{(127)^2}{6} + \frac{(44)^2}{6} + \frac{(89)^2}{6} + \frac{(40)^2}{6}\right) - 3(24+1) = 16.953$$

Since $I = 4$ groups, the sampling distribution of H is approximately chi-square with $4 - 1 = 3$ degrees of freedom. From Table F we see that the P-value is approximately 0.001. There is strong evidence of a difference in the number of insects trapped between the four groups. As in the one-way ANOVA, nonparametric multiple comparisons or contrasts would be required to explore the group differences further.

The Minitab software gives the output following when doing the Kruskal-Wallis test. The medians, average ranks (in place of sums of ranks), H statistic, and P-value are given. The H statistic with an adjustment for ties in the ranks is also given.

Kruskal-Wallis Test

```
LEVEL     NOBS     MEDIAN     AVE. RANK
    1        6      46.50        21.2
    2        6      15.50         7.3
    3        6      34.50        14.8
    4        6      15.00         6.7
OVERALL     24                   12.5

H = 16.95  d.f. = 3  p = 0.001
H = 16.98  d.f. = 3  p = 0.001 (adjusted for ties)
```

CHAPTER 16

BOOTSTRAP METHODS AND PERMUTATION TESTS

NOTE: In those questions in this chapter for which the answer is obtained by resampling, your answer may differ slightly from the one we give. This is because samples are chosen at random, so the particular set of samples you select will differ from ours. Because of this, we will depart from the Guided and Complete Solutions format of the other chapters and just provide Sample Solutions, which will include details on how to use the S-PLUS software to carry out the calculations.

SECTION 16.1

OVERVIEW

To **bootstrap** a statistic (such as the sample mean), we draw hundreds of **resamples** with replacement from the original data, calculate the statistic for each resample, and inspect the **bootstrap distribution** of the resampled statistics. This bootstrap distribution approximates the sampling distribution of the statistic. Notice that we use a quantity (the bootstrap distribution) based on the sample to approximate a similar quantity (the sampling distribution) from the population; this is called the **plug-in principle.**

In most cases, bootstrap distributions have approximately the same shape and spread as the sampling distribution. However, they are centered at the statistic computed from the original data. This is in contrast to the sampling distribution, which is centered at the parameter.

We use graphs and numerical summaries to determine whether the bootstrap distribution is approximately normal and is centered at the original statistic and to estimate the spread. The **bootstrap standard error** is the standard deviation of the bootstrap distribution.

The bootstrap is used to estimate the variation in a statistic based on the original data. It does not add to or replace the original data.

SAMPLE SOLUTIONS

Exercise 16.1

(a) and (b) The 20 resamples are given below, along with their means.

	3.67	17.30	17.30	5.32	0.00	29.78	3.67	3.67	17.30	0.00	26.47	3.67	26.47	0.00	17.30	17.30	17.30	3.67	26.47	17.30
	3.67	17.30	5.32	3.67	0.00	17.30	5.32	26.47	29.78	3.67	3.67	5.32	3.67	26.47	3.67	29.78	5.32	26.47	26.47	0.00
	3.67	3.67	17.30	0.00	5.32	29.78	0.00	3.67	29.78	17.30	5.32	17.30	26.47	3.67	29.78	0.00	3.67	17.30	26.47	17.30
	17.30	17.30	3.67	0.00	17.30	29.78	3.67	17.30	0.00	5.32	29.78	29.78	5.32	3.67	5.32	5.32	29.78	3.67	17.30	5.32
	3.67	0.00	3.67	5.32	0.00	17.30	29.78	29.78	5.32	3.67	3.67	29.78	3.67	0.00	3.67	29.78	0.00	0.00	5.32	3.67
	17.30	17.30	17.30	26.47	0.00	26.47	17.30	0.00	17.30	17.30	17.30	5.32	5.32	3.67	17.30	3.67	3.67	3.67	3.67	29.78
Mean	8.21	12.15	10.76	6.80	3.77	25.07	9.96	13.48	16.58	7.88	14.37	15.20	11.82	6.25	12.84	14.31	9.96	9.13	17.62	12.23

Your 20 resamples and the resulting means will differ from those given here.

(c) A stemplot of the means (after rounding down to the nearest integer) is given below. The stems have been split as well.

```
0  3
0
0  667
0  8999
1  01
1  2223
1  445
1  67
1
2
2
2  5
```

Because of the small number of resamples, your stemplot may look quite different from the one presented here.

(d) The bootstrap standard error is the standard deviation of the 20 means. Using software, we find that this is 4.73.

Exercise 16.10

Once you have loaded the resample library, import the data from the file *ex16_010.dat*.
(a) The S-PLUS command needed to produce this histogram is

```
> hist(ex16.010$days)
```

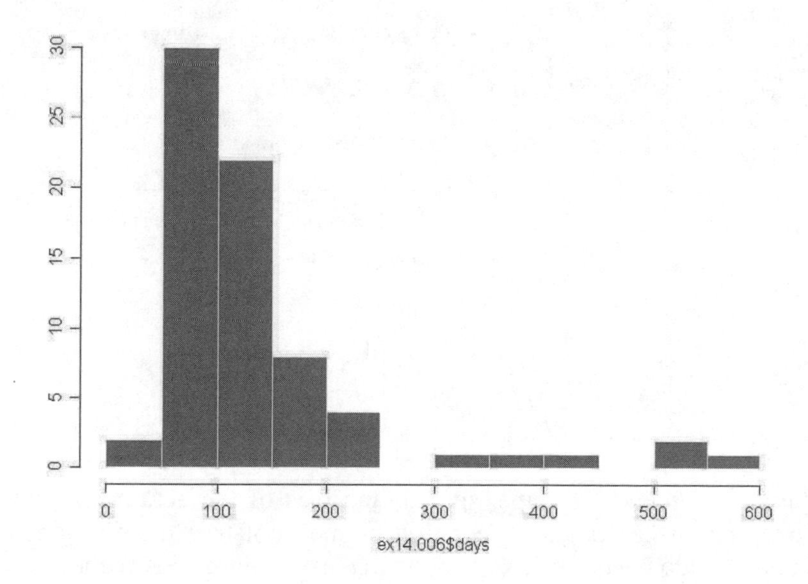

Notice that the histogram is strongly right-skewed with some high outliers.

(b) In the menu, go under Statistics to resample to bootstrap. In the window that opens, make sure the data set is "ex16.010," and in the expression type "mean(ex16.010$days)." Under the Results tab, make sure "Both Distribution and QQ" is checked under Plots. The bootstrap distribution and a normal quantile plot are given below. We see that the central part of the distribution is symmetric and close to normal. The quantile plot shows that the small and large means are both greater than would be expected under the normal distribution. This is caused by the high outliers in the original population. While this would not be a concern in raw data, here it occurs in a bootstrap distribution, after the central limit theorem has had a chance to work. Later in the chapter we learn ways to get more accurate confidence intervals in cases like this.

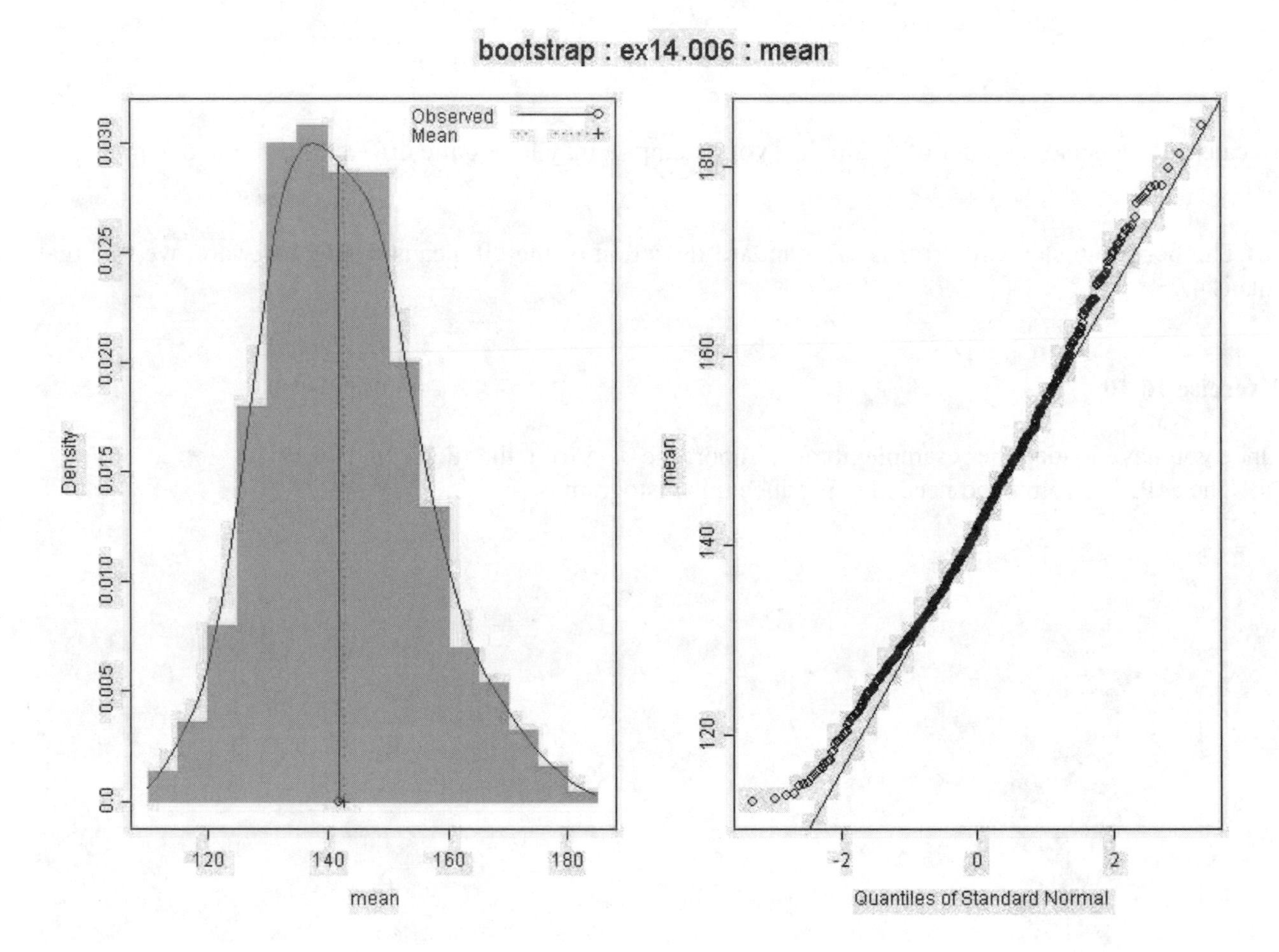

SECTION 16.2

OVERVIEW

The bootstrap distribution of most statistics mimics the shape, spread, and bias of the actual sampling distribution. The **bootstrap standard error** is the standard deviation of the bootstrap distribution. It measures how much a statistic varies under random sampling. The bootstrap estimate of **bias** is the mean of the bootstrap distribution minus the value of the statistic for the original data. The bias is small when the bootstrap distribution is centered at the value of the statistic for the original data. Small bias suggests that the sampling distribution of the statistic is centered at the value of the population parameter.

The bootstrap can estimate the sampling distribution, bias, and standard error of many statistics. One example is the trimmed mean.

If the bootstrap distribution is approximately normal (as seen in a histogram or quantile plot) and the bias is small, we can compute a **bootstrap *t* confidence interval** of the form

$$\text{statistic} \pm t^*\text{SE}$$

for the parameter. Do not use this *t* interval if the bootstrap distribution is not normal or if it shows substantial bias.

SAMPLE SOLUTIONS

Exercise 16.17

(a) Once you have loaded the resample library, import the data from the file *ex16_010.dat*. In the menu, go under Statistics to resample to bootstrap. In the window that opens, make sure the data set is "ex16.010," and in the expression box type "mean(ex16.010$days)." Under the Results tab, make sure "Both Distribution and QQ" is checked under Plots. The distribution and the QQ plot were included with Exercise 16.10 of this Study Guide. The additional bootstrap output is given below.

```
        *** Bootstrap Results ***
Call:
bootstrap(data = ex16.010, statistic = mean(ex16.010$days))

Number of Replications: 1000

Summary Statistics:
      Observed   Mean      Bias     SE
Param    141.8 141.3 -0.5164 12.83
```

The bias is –0.5164. This is small compared to the observed mean of 141.8. The bootstrap distribution of most statistics (such as the sample mean) mimics the shape, spread, and bias of the actual sampling distribution. Thus, we expect that the bias encountered using $\bar{x}$ to estimate the mean survival time for all guinea pigs that receive the same experimental treatment is also small. The use of the bootstrap *t* confidence interval is justified.

(b) The general formula for the bootstrap *t* confidence interval is

$$\text{statistic} \pm t^*\text{SE}$$

In this example, the value of the statistic is the sample mean, which is 141.8, the value of t^* with $n - 1 =$ 71 degrees of freedom is 1.99, and SE = 12.83. This gives a bootstrap *t* confidence interval of

$$141.8 \pm (1.99)(12.83) = (116.27, 167.33)$$

(c) The numerical value of SE_{boot} = 12.83 and the value for the formula based standard error is $s/\sqrt{n} = 109.2/\sqrt{72} = 12.9$, which are in very close agreement. The usual *t* 95% confidence interval is

$$\bar{x} \pm t^* s/\sqrt{n} = 141.8 \pm (1.99)(12.9) = (116.2, 167.5)$$

and we see the two intervals are in very close agreement.

Exercise 16.21

Once you have loaded the resample library, import the data from the file *ex16_021.dat.*
(a) The S-PLUS commands needed to produce the histogram and quantile plot are

```
> hist(ex16.021$value)
> qqnorm(ex16.021$value)
```

The histogram and normal quantile plot of the data are given below. There do not appear to be any significant departures from normality. The histogram is centered at about 0, the spread is approximately what we would expect for the $N(0, 1)$ distribution, and the points in the normal quantile plot follow a straight line.

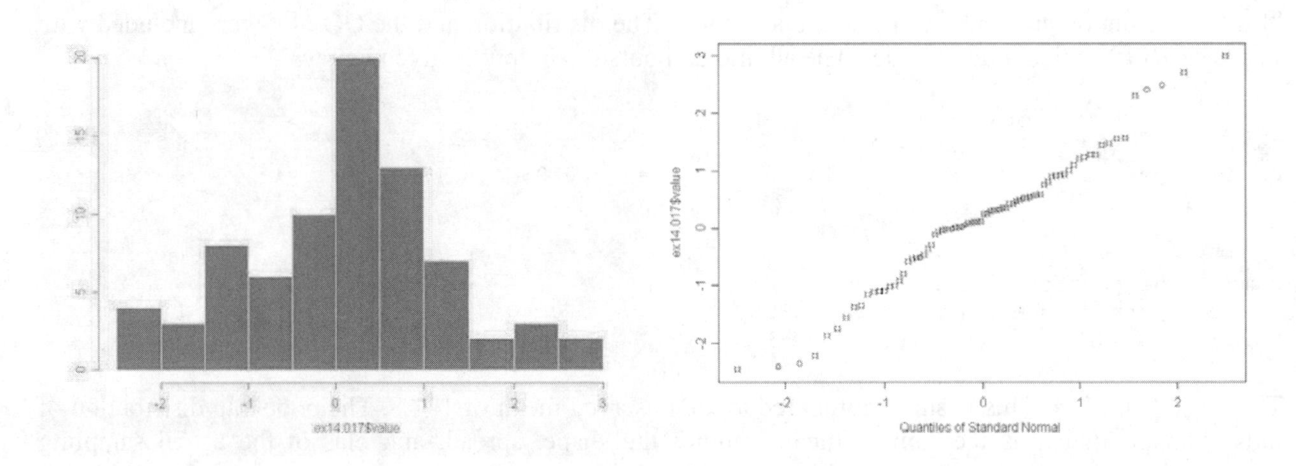

(b) In the menu, go under Statistics to resample to bootstrap. In the window that opens, make sure the data set is "ex16.021," and in the expression box type "mean(ex16.021$value)." Under the Results tab, make sure "Both Distribution and QQ" is checked under Plots. The bootstrap output is given below.

```
        *** Bootstrap Results ***
Call:
bootstrap(data = ex16.021, statistic = mean(ex16.021$value))

Number of Replications: 1000

Summary Statistics:
       Observed   Mean     Bias    SE
Param    0.1249 0.1308 0.005953 0.128
```

We see that the standard error of the bootstrap mean is 0.128.

The bootstrap distribution and a normal quantile plot are given on the following page.

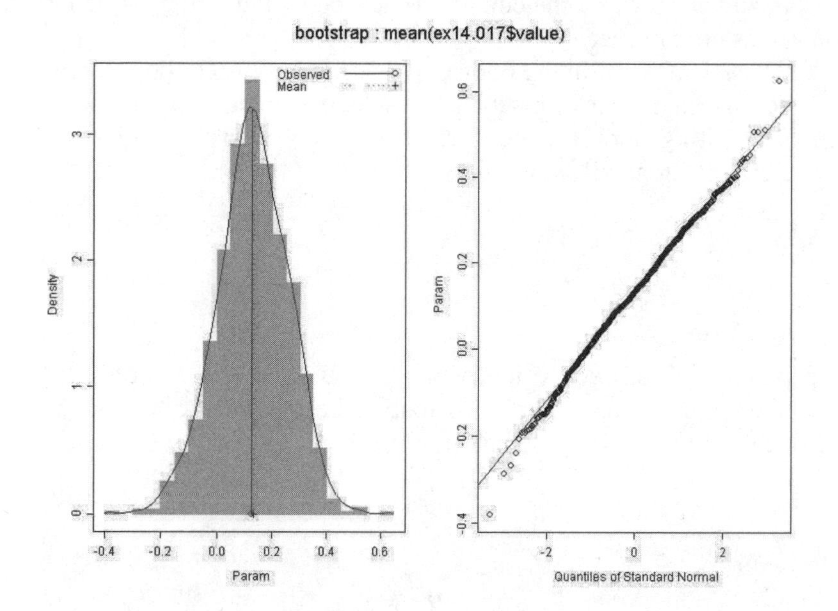

The bias is small 0.005953 and the histogram of the data in part (a) looks approximately normal, so that the bootstrap distribution of the mean will also look normal. The bootstrap distribution of the mean given above confirms this. Neither the histogram nor the quantile plot of the mean shows significant departures from normality. When the bootstrap distribution is approximately normal and the bias is small, it is safe to use the bootstrap t confidence interval.

(c) From part (b), we see that the bootstrap mean is 0.1308 and the standard error is SE = 0.128. The sample size is $n = 78$, and the upper 0.025 percentile of the t distribution with $n - 1 = 77$ df is found, using Table D (use the value that corresponds to 80 df), to be $t^* = 1.99$. Thus,

$$\text{bootstrap mean} \pm t^*\text{SE} = 0.1308 \pm (1.99)(0.128) = 0.1308 \pm 0.2547 = (\ 0.1239, 0.3855)$$

The formula-based interval is

$$\bar{x} \pm t^* \frac{s}{\sqrt{n}} = 0.1249 \pm 1.99\left(\frac{1.1556}{\sqrt{78}}\right) = 0.1249 \pm 0.2607 = (-0.1358, 0.3856)$$

which is in very good agreement with the bootstrap method.

SECTION 16.4

OVERVIEW

Both bootstrap t and traditional z and t confidence intervals require statistics with small biases and sampling distributions close to normal. We can check these conditions by examining the bootstrap distribution (a histogram of the bootstrap, a quantile plot, and summary statistics) for bias and lack of normality.

The **bootstrap percentile confidence interval** for 95% confidence is the interval between the 2.5% percentile and the 97.5% percentile of the bootstrap distribution. Agreement between the bootstrap t and percentile intervals is an added check on the conditions needed by the t interval. Do not use t or percentile intervals if these conditions are not met.

When bias or skewness is present in the bootstrap distribution, use either a **bootstrap tilting** or **BCa** interval. The t and percentile interval give inaccurate results under these circumstances unless the sample sizes are very large. The tilting and BCa confidence intervals adjust for bias and skewness and are generally very accurate except for small samples.

SAMPLE SOLUTIONS

Exercise 16.30

For a 90% bootstrap percentile confidence interval, we would use the 5% and 95% percentiles as the endpoints. This is because the percent of the distribution between the 5% and the 95% percentiles is 95% – 5% = 90%. For the 98% bootstrap percentile confidence interval, we would use the 1% and 99% percentiles as the endpoints.

Exercise 16.39

The sample mean $\overline{x}$ = 141.847, and software gives the formula-based one-sample t interval as

$$\overline{x} \pm t^* \frac{s}{\sqrt{n}} = (116.184, 167.510)$$

In the menu, go under Statistics to resample to bootstrap. In the window that opens, make sure the data set is "ex16.010," and in the expression box type "mean(ex16.010$days)." Under the Results tab, make sure that "Tilting Confidence Intervals" and "t Intervals using Bootstrap SE" are both checked. The resulting output is given below.

```
          *** Bootstrap Results ***
Call:
bootstrap(data = ex16.010, statistic = mean(ex16.010$days), L = "choose")

Number of Replications: 1000

Summary Statistics:
      Observed  Mean    Bias    SE
Param    141.8 141.7 -0.1327 12.43

Percentiles:
          2.5%       5%      95%    97.5%
Param 118.5573 122.2535 162.2361 169.0833

BCa Confidence Intervals:
          2.5%       5%      95%    97.5%
Param 121.0315 123.6322 166.3922 172.3548

Tilting Confidence Intervals:
          2.5%       5%      95%    97.5%
Param 121.0443 123.9918 167.4633 173.2733

T Confidence Intervals using Bootstrap Standard Errors:
          2.5%       5%      95%   97.5%
Param 117.0304 121.1107 162.5838 166.664
```

In the graph below, the four intervals are displayed as horizontal lines, one above the other. A vertical line is drawn at the original sample mean $\overline{x} = 141.8$. Both the BCa interval and the tilting interval are not symmetric about the mean. The intervals are shorter below the mean than they are above the mean. The bootstrap *t* interval is fairly symmetric about the mean $\overline{x}$, and the percentile interval slightly less so.

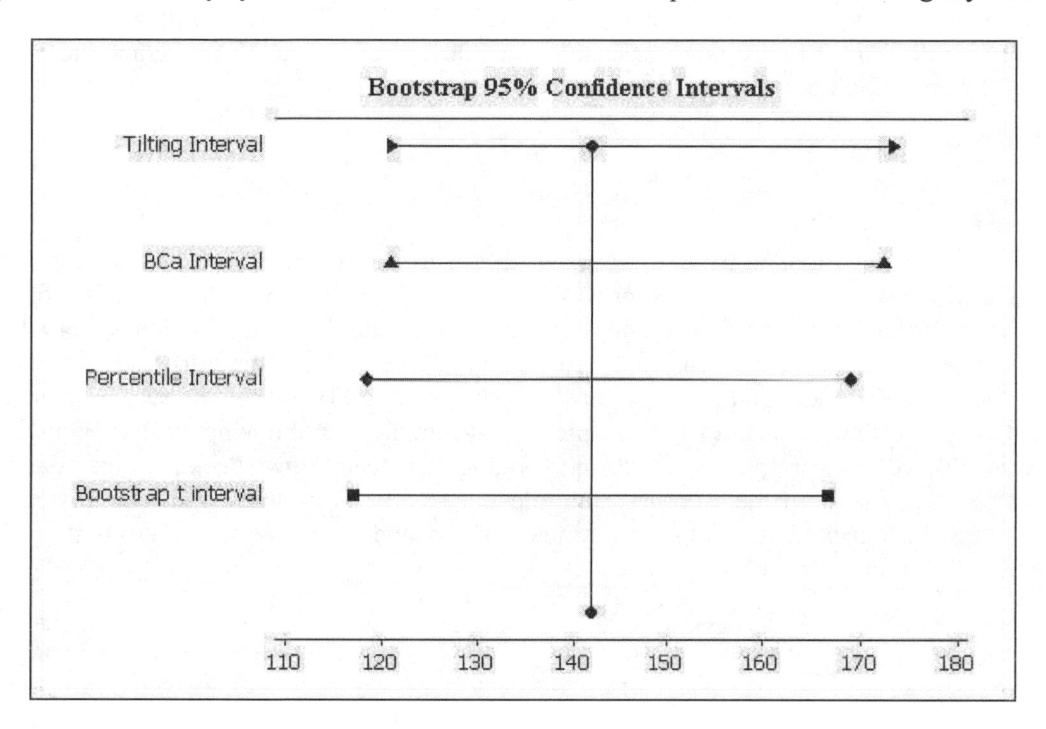

SECTION 16.5

OVERVIEW

Permutation tests are significance tests based on **permutation resamples** drawn at random from the original data. Permutation resamples are drawn **without replacement,** in contrast to bootstrap samples, which are drawn with replacement. Permutation resamples must be drawn in a way that is consistent with the null hypothesis and with the study design. In a **two-sample design,** the null hypothesis says that the two populations are identical. Resampling randomly assigns observations to the two groups. In a **matched pairs** design, randomly permute the two observations within each pair separately. To test the hypothesis of **no relationship** between two variables, randomly reassign values of one of the two variables.

The **permutation distribution** of a suitable statistic is formed by the values of the statistic in a large number of resamples. Find the *P*-value of the test by locating the original value of the statistic on the permutation distribution.

When they can be used, permutation tests have great advantages. They do not require specific population shapes such as normality. They apply to a variety of statistics, not just to statistics that have a simple distribution under the null hypothesis. They can give very accurate *P*-values, regardless of the shape and size of the population (if enough permutations are used).

It is often useful to give a confidence interval along with a test. To create a confidence interval, we no longer assume the null hypothesis is true, so we use bootstrap resampling rather than permutation resampling.

SAMPLE SOLUTIONS

Exercise 16.54

(a) Let μ_1 denote the mean selling price for all Seattle real estate transactions in 2002 and μ_2 the mean selling price for all Seattle real estate transactions in 2001. We wish to test whether these means are significantly different. To do this, we test the hypotheses

$$H_0: \mu_1 = \mu_2$$
$$H_a: \mu_1 \neq \mu_2$$

(b) To perform the calculations in parts (b), (c), and (d), we use S-PLUS. The data set should contain two columns. The first column should be the price. There should be 100 rows of data. The first 50 rows should be the 50 values from Table 16.1, and the next 50 rows should be the 50 values from Table 16.5. The second column should be the year, and will also have 100 rows. The entries in the first 50 rows should be "2002," and the next 50 rows should contain the entry "2001." The data set is named *ex16.054*.

In the menu, go under Statistics to resample to two-sample *t*. In the window that opens, make sure the data set is "ex16.054," variable 1 is Price, and variable 2 is Year. Now check the box "Variable 2 is a grouping variable." Click on the Bootstrap tab and check "Perform Bootstrap." Make sure that BCa confidence interval is checked. Click on the Permuations tab, and check "Perform Permutation Test."

The S-PLUS output for the two-sample *t* test is

```
      Welch Modified Two-Sample t-Test

data:   x: Price with Year = 2002 , and y: Price with Year = 2001
t = 0.8057, df = 71.8952170780399, p-value = 0.4231
alternative hypothesis:  difference in means is not equal to 0
95 percent confidence interval:
  -59.45468  140.11588
sample estimates:
 mean of x mean of y
  329.2571  288.9265
```

To carry out a significance test for $H_0: \mu_1 = \mu_2$, recall that we use the two-sample t statistic

$$t = \frac{(\bar{x}_1 - \bar{x}_2)}{\sqrt{\frac{s_1^2}{n_1} + \frac{s_2^2}{n_2}}}$$

We know that $n_1 = n_2 = 50$, and in the S-PLUS notation x corresponds to the first group (year = 2002) and y to the second group (year = 2001). From the S-PLUS output we find

$$\bar{x}_1 = 329.2571 \text{ and } \bar{x}_2 = 288.9265$$

and we can calculate $s_1 = 316.83$ and $s_2 = 157.7778$. Thus,

$$t = \frac{329.2571 - 288.9265}{\sqrt{\frac{(316.83)^2}{50} + \frac{(157.7778)^2}{50}}} = \frac{40.3306}{\sqrt{2505.5017}} = 0.8057$$

as in the S-PLUS output. The P-value is given as 0.4231, which uses the estimated degrees of freedom 71.895. There is little evidence that the population means μ_1 and μ_2 differ.

(c) The S-PLUS output for the permutation test is

```
         *** Permutation Test Results ***
Call:
permutationTestMeans(data = ex16.054$Price, treatment = ex18.049$Year, B = 999)

Number of Replications: 999

Summary Statistics:
    Observed Mean    SE alternative p.value
Var    40.33  2.4 49.17   two.sided   0.458
```

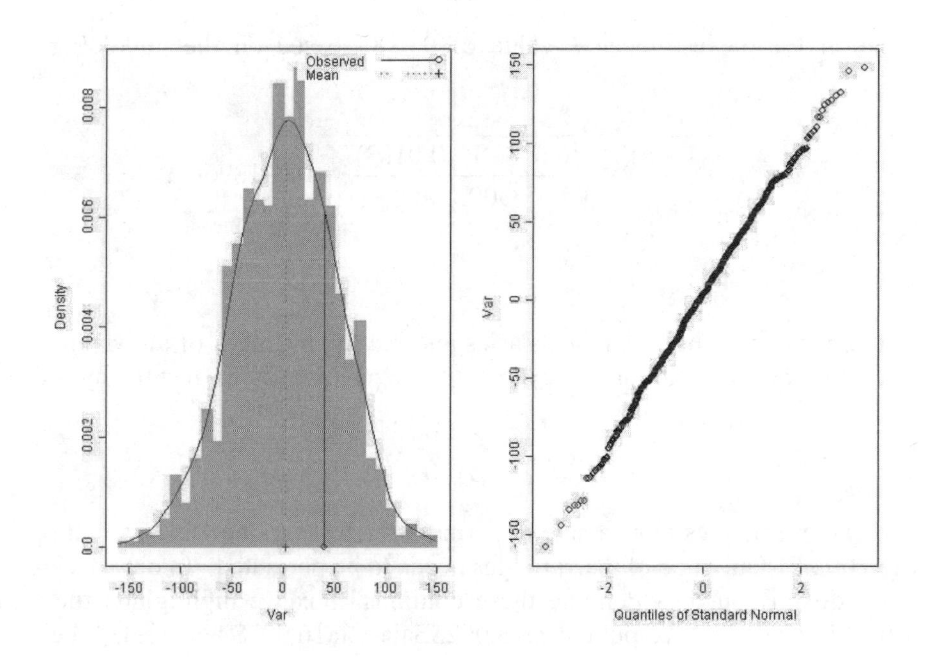

From the output, we see that the P-value is 0.458. This is consistent with the P-value we computed in (b). We again conclude that there is little evidence that the population means μ_1 and μ_2 differ.

(d) The S-PLUS output for the BCa interval is

```
BCa Confidence Intervals:
            2.5%         5%       95%     97.5%
Param  -35.87395  -24.53738  154.2974  175.5956
```

A 95% BCa confidence interval for the mean change from 2001 to 2002 is obtained by using the 2.5% and 97.5% percentiles as the lower and upper confidence limits. The interval is

$$(-35.87395, 175.5956)$$

This interval includes 0 and suggests that the two means are not significantly different at the 0.05 level. Again, this is consistent with the conclusions in parts (b) and (c).

Exercise 16.56

The guideline given in the chapter states: "If the true (one-sided) P-value is p, the standard deviation of the estimated P-value is approximately

$$\sqrt{\frac{p(1-p)}{B}}$$

where B is the number of resamples." Using this guideline we find the following.

The standard deviation for the estimated P-value of 0.015 for the DRP study, based on B = 999 resamples, is

$$\sqrt{\frac{p(1-p)}{B}} = \sqrt{\frac{0.015(1-0.015)}{999}} = 0.003846$$

The standard deviation for the estimated P-value of 0.0183 based on the 500,000 resamples in the Verizon study is

$$\sqrt{\frac{p(1-p)}{B}} = \sqrt{\frac{0.0183(1-0.0183)}{500,000}} = 0.001896$$

Exercise 16.62

(a) Let ρ denote the correlation between the salaries and batting averages of all Major League Baseball players. In this case, if there is correlation, we expect it to be positive, so we test the hypotheses

$$H_0: \rho = 0$$
$$H_a: \rho > 0$$

(b) In the menu, go under Statistics to resample to permutation test. In the window that opens, make sure the data set is "Ta16.002." Only one of the variables needs to be permuted. In our case, we permuted the variable Salary. To do this, in the "Permute these columns" box we highlighted the variable Salary. Then, in the box for "Expression" we put cor(ta16.002$Salary,ta16.002$Average). The resulting output is given below.

```
Call:
permutationTest(data = ta16.002, statistic = cor(ta14.002$Salary, ta16.002$
      Average), alternative = "two.sided", resampleColumns = "Salary")

Number of Replications: 999

Summary Statistics:
      Observed      Mean     SE alternative p-value
Param   0.1068 -0.001748 0.1469   two.sided   0.448
```

We want the one-sided P-value, so we divide the two-sided value given in the output by 2. We then see that the P-value is

$$P\text{-value} = 0.224$$

and we conclude that there is not strong evidence that salaries and batting averages are correlated in the population of all major league players.

CHAPTER 17

STATISTICS FOR QUALITY: CONTROL AND CAPABILITY

SECTION 17.1

OVERVIEW

In practice, work is often organized into a chain of activities that lead to some result. Such a chain of activities that turns inputs into outputs is called a **process.** A process can be described by a **flowchart,** which is a picture of the stages of a process. A **cause-and-effect diagram,** which displays the logical relationships between the inputs and outputs of a process, is also useful for describing and understanding a process.

All processes have variation. If the pattern of variation is stable over time, the process is said to be in statistical control. In this case, the sources of variation are called **common causes.** If the pattern is disrupted by some unusual event, **special cause** variation is added to the common cause variation. **Control charts** are statistical plots intended to warn when a process is disrupted or **out of control.**

Standard **3σ control charts** plot the values of some statistic Q for regular samples from the process against the time order in which the samples were collected. The **center line** of the chart is at the mean of Q. The **control limits** lie three standard deviations of Q above (the **upper control limit**) and below (the **lower control limit**) the center line. A point outside the control limits is an **out-of-control signal.** For **process monitoring** of a process that has been in control, the mean and standard deviations used to establish the center line and control limits are based on past data and are updated regularly.

When we measure some quantitative characteristic of a process, we use $\bar{x}$ **and** s **charts** for process control. The $\bar{x}$ chart plots the sample means of samples of size n from the process, and the s chart plots the sample standard deviations. The s chart monitors variation within individual samples from the process. If the s chart is in control, the $\bar{x}$ chart monitors variation from sample to sample. To interpret charts, always look first at the s chart.

For a process that is in control with mean μ and standard deviation σ, the 3σ $\bar{x}$ chart based on samples of size n has center line and control limits

$$\text{CL} = \mu,\ \text{UCL} = \mu + 3\frac{\sigma}{\sqrt{n}},\ \text{LCL} = \mu - 3\frac{\sigma}{\sqrt{n}}$$

The 3σ s chart has control limits

$$\text{UCL} = (c_4 + 3c_5)\sigma = B_6\sigma, \text{LCL} = c_4 - 3c_5)\sigma = B_5\sigma$$

and the values of c_4, c_5, B_5, and B_6 can be found in Table 17.2 in the text for values of n from 2 to 10.

GUIDED SOLUTIONS

Exercise 17.1

KEY CONCEPTS: Flowchart and cause-and effect diagram

For this exercise, it is important to choose a process that you know well so that you can describe it carefully and recognize those factors that affect the process. We take as our example the process of making a good cup of coffee. Some of the stages in the process to be included in the flow chart are measuring the coffee, grinding the coffee, and so forth. In making a cause-and-effect diagram, consider what factors might affect the final cup of coffee at each stage of the flowchart.

Exercise 17.3

KEY CONCEPTS: Common cause variation, special causes

In Exercise 17.1, we described the process of making a good cup of coffee. Some sources of common cause variation are variation in how long the coffee has been stored and the conditions under which it has been stored. A special cause that might drive the process out of control could be the use of milk or cream that has gone bad. Now list some sources of common cause variation and special causes for the process that you described in Exercise 17.1.

Exercise 17.17

KEY CONCEPTS: Center line and control limits, $\bar{x}$ and s charts

We are told that $n = 5$, $\mu = 0.8750$, and $\sigma = 0.0012$. For the s chart, use Table 17.2 for the values of c_4, B_5, and B_6.

UCL = $B_6\sigma =$
CL = $c_4\sigma =$
LCL = $B_5\sigma =$

For the $\bar{x}$ chart,

$$\text{UCL} = \mu + 3\frac{\sigma}{\sqrt{n}} =$$

$$\text{CL} = \mu =$$

$$\text{LCL} = \mu - 3\frac{\sigma}{\sqrt{n}} =$$

Exercise 17.19

KEY CONCEPTS: $\bar{x}$ and s charts

First compute $\bar{x}$ and s for the first two samples:

first sample: $\bar{x} =$, $s =$; second sample: $\bar{x} =$, $s =$

To make the s chart, we note that

$$\text{UCL} = B_6\sigma = 1.964(12.74) = 25.02$$
$$\text{CL} = c_4\sigma = 0.9400(12.74) = 11.98$$
$$\text{LCL} = B_5\sigma = 0(12.74) = 0$$

resulting in the chart that follows.

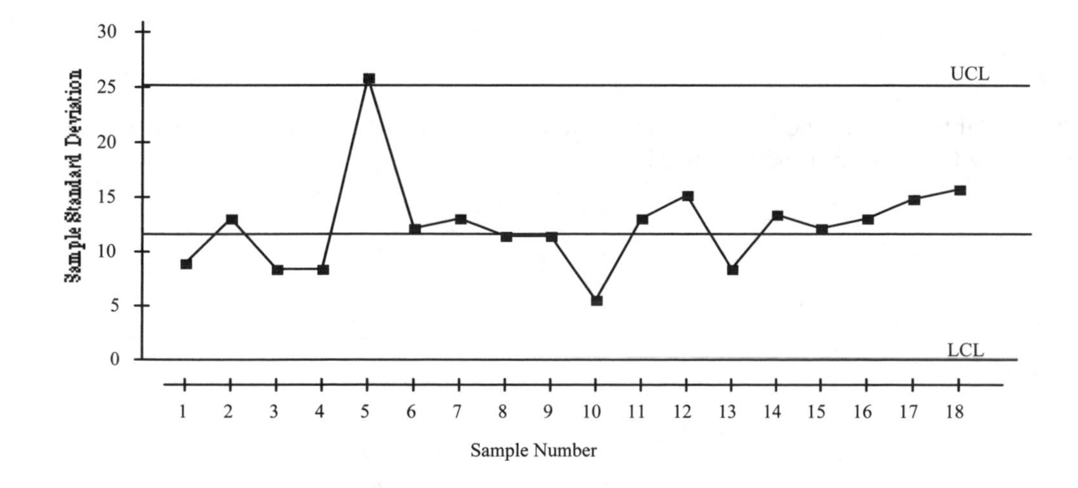

To make the $\bar{x}$ chart, first find the center line and the control limits.

UCL =

CL =

LCL =

Now plot the means, the center line, and the control limits on the axes below to give the $\bar{x}$ chart.

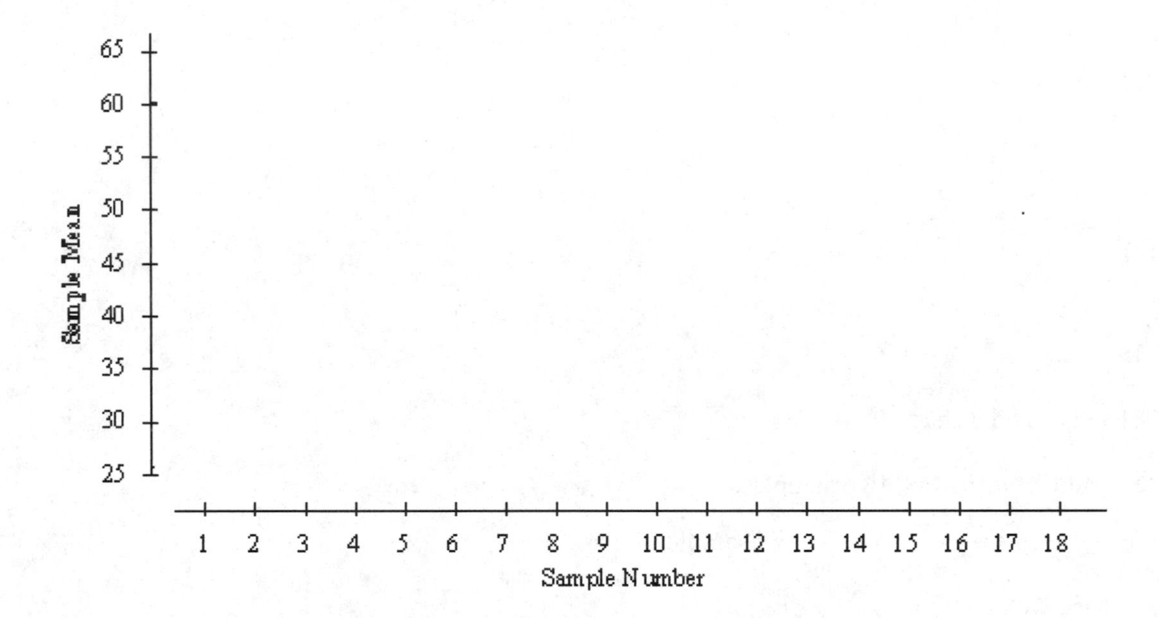

Now use these charts to describe the state of the process.

If you have software available, you should learn how to use it to create a control chart. The control charts in this chapter of the Study Guide were produced in Minitab.

COMPLETE SOLUTIONS

Exercise 17.1

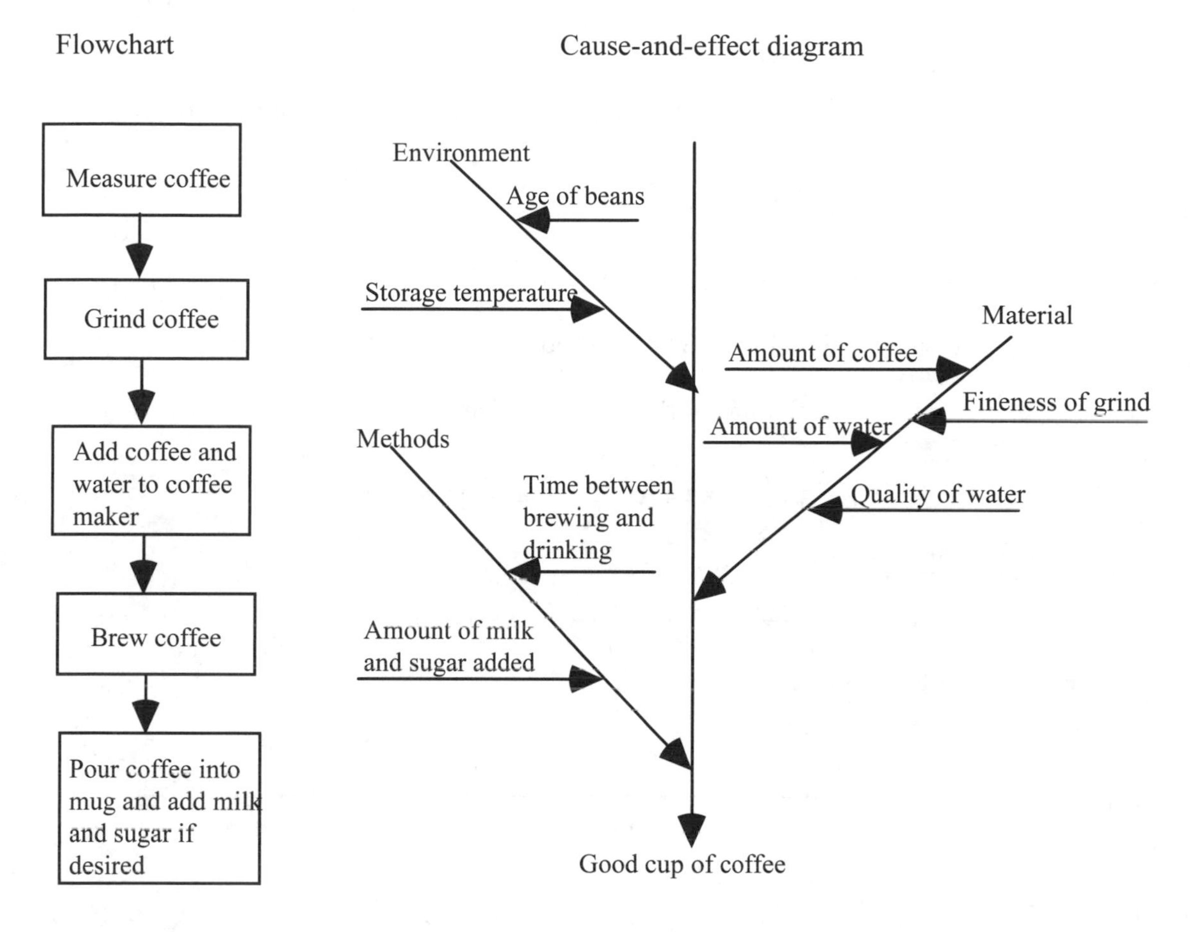

Exercise 17.3

In Exercise 17.1, we described the process of making a good cup of coffee. Some sources of common cause variation are variation in how long the coffee has been stored and the conditions under which it has been stored, variation in the measured amount of coffee used, variation in how finely ground the coffee is, variation in the amount of water added to the coffeemaker, variation in the length of time the coffee sits between when it has finished brewing and when it is drunk, and variation in the amount of milk and/or sugar added.

Some special causes that might at times drive the process out of control would be a bad batch of coffee beans, a serious mismeasurement of the amount of coffee used or the amount of water used, a malfunction of the coffee maker or a power outage, interruptions that result in the coffee sitting a long time before it is drunk, and the use of milk that has gone bad.

Exercise 17.17

For the s chart, using Table 17.2 for the values of c_4, B_5, and B_6 we have

$$\text{UCL} = B_6\sigma = 1.964(0.0012) - 0.0023568$$
$$\text{CL} = c_4\sigma = 0.9400(0.0012) = 0.001128$$
$$\text{LCL} = B_5\sigma = 0(0.0012) = 0$$

For the $\bar{x}$ chart,

$$\text{UCL} = \mu + 3\frac{\sigma}{\sqrt{n}} = 0.8750 + 3\frac{0.0012}{\sqrt{5}} = 0.8750 + 3(0.00054) = 0.87662$$
$$\text{CL} = \mu = 0.8750$$
$$\text{LCL} = \mu - 3\frac{\sigma}{\sqrt{n}} = 0.8750 - 3\frac{0.0012}{\sqrt{5}} = 0.8750 - 3(0.00054) = 0.87338$$

Exercise 17.19

We compute $\bar{x}$ and s for the first two samples and find

first sample: $\bar{x} = 48$, $s = 8.94$; second sample: $\bar{x} = 46$, $s = 13.03$

The s chart is given in the Guided Solutions. To make the $\bar{x}$ chart, we note that

$$\text{UCL} = \mu + 3\frac{\sigma}{\sqrt{n}} = 43 + 3\frac{12.74}{\sqrt{5}} = 43 + 17.09 = 60.09$$
$$\text{CL} = \mu = 43$$
$$\text{LCL} = \mu - 3\frac{\sigma}{\sqrt{n}} = 43 - 3\frac{12.74}{\sqrt{5}} = 43 - 17.09 = 25.91$$

resulting in the chart that follows.

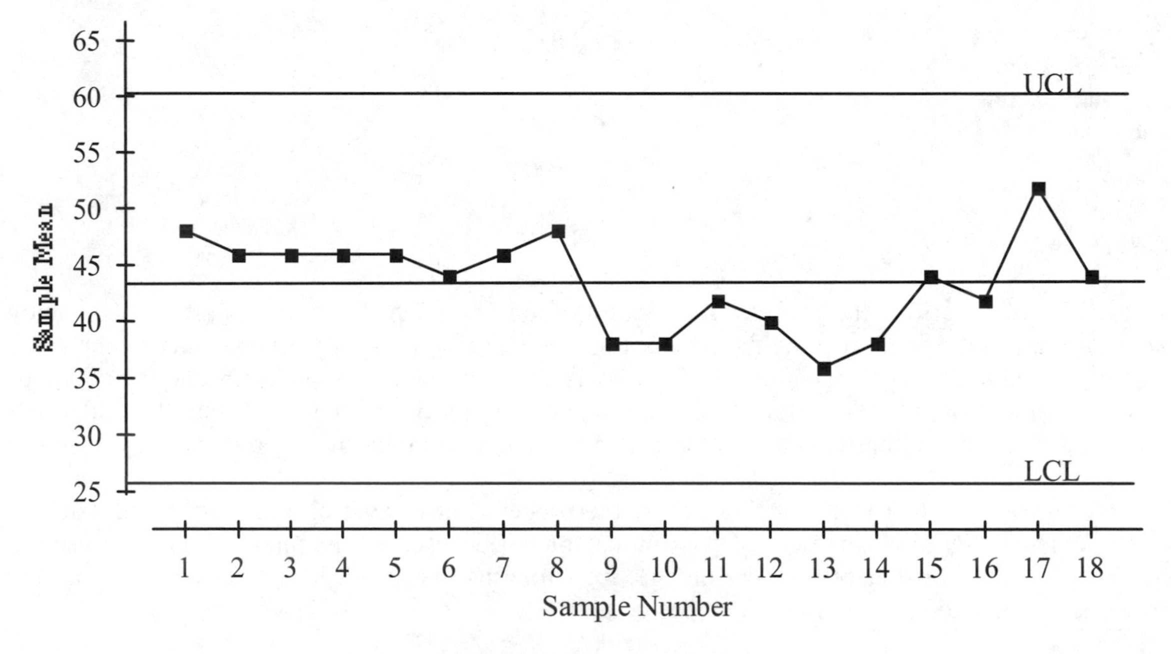

The s chart shows a lack of control at sample point 5 (8:15 3/8), but otherwise neither chart shows a lack of control. We would want to find out what happened at sample 5 to cause a lack of control in the s chart.

SECTION 17.2

OVERVIEW

An ***R* chart** based on the range of observations in a sample is often used in place of an s chart. We will rely on software to produce such charts. Formulas can be found in books on quality control. $\bar{x}$ and R charts are interpreted the same way as $\bar{x}$ and s charts.

It is common to use various **out-of-control signals** in addition to "one point outside the control limits." In particular, a **runs signal** (nine consecutive points above the center line or nine consecutive points below the center line) for an $\bar{x}$ chart allows one to respond more quickly to a gradual drift in the process center.

We almost never know the mean μ and standard deviation σ of a process. These must be estimated from past data. We estimate μ by the mean $\bar{\bar{x}}$ of the observed sample means $\bar{x}$. We estimate σ by

$$\hat{\sigma} = \frac{\bar{s}}{c_4}$$

where $\bar{s}$ is the mean of the observed sample standard deviations. **Control charts based on past data** are used at the **chart setup** stage for a process that may not be in control. Start with control limits calculated from the same past data that you are plotting. Beginning with the s chart, narrow the limits as you find special causes and remove the points influenced by these causes. When the remaining points are in control, use the resulting limits to monitor the process.

Statistical process control maintains quality more economically than inspecting the final output of a process. Samples that are **rational subgroups** (subgroups that capture the features of the process in which we are interested) are important for effective control charts. A process in control is stable, so that we can predict its behavior. If individual measurements have a normal distribution, we can give the **natural tolerances.**

A process is **capable** if it can meet or exceed the requirements placed on it. Control (stability over time) does not in itself improve capability. Remember that control describes the internal state of the process, whereas capability relates the state of the process to external specifications.

GUIDED SOLUTIONS

Exercise 17.29

KEY CONCEPTS: $\bar{x}$ chart

The samples will be either the mean of times from the experienced clerk or the mean of times from the inexperienced clerk. Means of times from the experienced clerk will be relatively small and vary little from sample to sample. Means of times from the inexperienced clerk will be relatively large and vary little from sample to sample. Presumably, a given sample is equally likely to be from the experienced clerk or the inexperienced clerk, and which clerk a sample comes from will be random. Thus, the $\bar{x}$ chart should display two types of points: those that have relatively small values and those with relatively large values. Both types should occur about equally often in the chart, and the pattern of large and small values should appear random. Sketch an $\bar{x}$ chart pattern that will have these properties.

Exercise 17.31

KEY CONCEPTS: Estimating μ and σ from past data

(a) From the values of $\bar{x}$ and s in Table 17.1, compute (using software or a calculator)

$\bar{\bar{x}}$ = mean of the 20 values of $\bar{x}$ =

$\bar{s}$ = mean of the 20 values of s =

Your estimate of μ is

$\hat{\mu} = \bar{\bar{x}} = 275.065$

and the estimate of σ is

$$\hat{\sigma} = \frac{\bar{s}}{c_4} =$$

(b) Examine the s chart in Figure 17.7. Explain why the s chart suggests that σ may now be less than 43 mV.

Exercise 17.37

KEY CONCEPTS: Adjusting the process to improve capability

The new specifications are that the mesh tension should be between 150 and 350 mV. According to Exercise 17.36, the standard deviation of the process is 38.4 mV. If we adjust the process mean to be 250 mV, then (assuming mesh tension is normally distributed) mesh tensions will follow an $N(\mu = 250, \sigma = 38.4)$ distribution. The probability that a monitor has a mesh tension x that is within the new specifications is

$P(150 < x < 350)$

Exercise 17.39

KEY CONCEPTS: Natural tolerances

The natural tolerances are $\mu \pm 3\sigma$. We do not know μ and σ, so we must estimate them from the data. We remove sample 5 from the data. Based on the remaining 17 samples, we estimate μ by the mean of the 17 values of $\bar{x}$, while σ should be estimated from all 85 remaining observations rather than by averaging the 17 within-sample standard deviations.

$\bar{\bar{x}}$ = mean of the 17 values of $\bar{x}$ =

$s =$ standard deviation of the remaining 85 observations =

Thus, we estimate μ to be

$\hat{\mu} = \bar{\bar{x}} =$

and we estimate σ to be (using the fact that c_4 is very close to 1 for a sample of size 85)

$\hat{\sigma} = s =$

Based on these estimates, the natural tolerances for the distance between the holes are

$\hat{\mu} \pm 3\hat{\sigma} =$

Exercise 17.41

KEY CONCEPTS: Percent meeting specifications

Based on the 17 samples that were in control, we saw in Exercise 17.39 that estimates of μ and σ are $\hat{\mu}$ = 43.41 and $\hat{\sigma}$ = 11.58. We therefore assume that distances between holes vary from meter to meter according to an $N(43.41,11.58)$ distribution. The probability that the distance x between holes in a randomly selected meter is between 54 ± 10 (i.e., between 44 and 64) is thus

$P(44 < x < 64)$ =

Exercise 17.49

KEY CONCEPTS: Process variation, $\bar{x}$ chart

The winning time for the Boston Marathon has been gradually decreasing over time. The process has not become stable but instead has steadily drifted downward. Thus, two sources of variation are in the data. One is the process variation that we would observe if times were stable (random variation that we might observe over short periods), and the other is variation due to the downward trend in times (the large differences between times from the early 1950s and recent times). There is a problem in using s to estimate the process variation. Do you think s will underestimate or overestimate the process variation? How will this affect the resulting control limits?

Complete Solutions

Exercise 17.29

The $\bar{x}$ chart below displays two types of points: those that have relatively small values (experienced clerk) and those with relatively large values (inexperienced clerk). Both types occur about equally often in the chart, and the pattern of large and small values appears random. We have not included control limits on the chart because the purpose is merely to illustrate the pattern of points that we might expect to see.

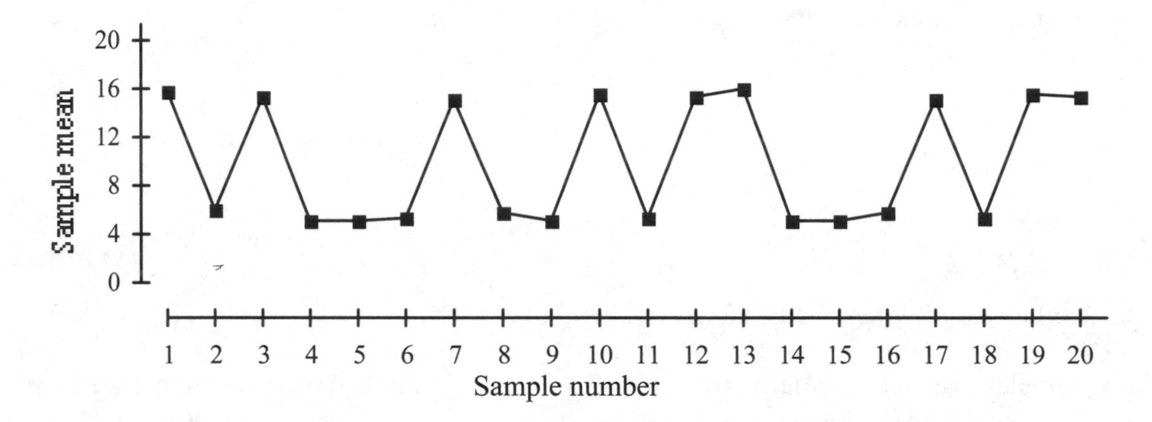

Exercise 17.31

(a) From the values of $\bar{x}$ and s in Table 12.1, we compute (using software)

$$\bar{\bar{x}} = \text{mean of the 20 values of } \bar{x} = 275.065$$
$$\bar{s} = \text{mean of the 20 values of } s = 34.55$$

hence, we estimate μ to be

$$\hat{\mu} = \bar{\bar{x}} = 275.065$$

and we estimate σ to be (using the fact that the samples are each of size $n = 4$ and according to Table 12.3, $c_4 = 0.9213$)

$$\hat{\sigma} = \frac{\bar{s}}{c_4} = \frac{34.55}{0.9213} = 37.5$$

(b) If we look at the s chart in Figure 12.7 we see that most of the points lie below 40 (and more than half of those below 40 lie well below 40), while of the points above 40, all but one (sample 12) are only slightly larger than 40. The s chart suggests that typical values of s are below 40, which is consistent with the estimate of σ in part (a).

Exercise 17.37

The probability a monitor has a mesh tension x that is within the new specifications is

$$P(150 < x < 350) = P\left(\frac{150-250}{38.4} < \frac{x-250}{38.4} < \frac{350-250}{38.4}\right) = P(-2.60 < Z < 2.60)$$

$$= P(Z < 2.60) - P(Z < -2.60) = 0.9953 - 0.0047 = 0.9906$$

where we have used Table A to compute $P(Z < 2.60)$ and $P(Z < -2.60)$. Thus, 99.06% of monitors will meet the new specifications if the process is adjusted to have center 250 mV.

Exercise 17.39

The natural tolerances are $\mu \pm 3\sigma$. We do not know μ and σ, so we must estimate them from the data. We remove sample 5 from the data. Based on the remaining 17 samples, we find

$$\bar{\bar{x}} = \text{mean of the 17 values of } \bar{x} = 43.41$$

$$s = \text{standard deviation of the remaining 85 observations} = 11.58$$

hence, we estimate μ to be

$$\hat{\mu} = \bar{\bar{x}} = 43.41$$

and we estimate σ to be (using the fact that c_4 is very close to 1 for a sample of size 85)

$$\hat{\sigma} = s = 11.58$$

Based on these estimates, the natural tolerances for the distance between the holes are

$$\hat{\mu} \pm 3\hat{\sigma} = 43.41 \pm 3(11.58) = 43.41 \pm 34.74 \text{ or } 8.67 \text{ to } 78.15.$$

Exercise 17.41

We assume that distances between holes vary from meter to meter according to an $N(43.41, 11.58)$ distribution. The probability that the distance x between holes in a randomly selected meter is between 54 $\pm$ 10 (i.e., between 44 and 64) is thus

$$P(44 < x < 64) = P\left(\frac{44 - 43.41}{11.58} < \frac{x - 43.41}{11.58} < \frac{64 - 43.41}{11.58}\right) = P(0.05 < Z < 1.78)$$

$$= P(Z < 1.78) - P(Z < 0.05) = 0.9625 - 0.5199 = 0.4426$$

We conclude that about 44.26% of meters meet specifications.

Exercise 17.49

The winning time for the Boston Marathon has been gradually decreasing over time. The process has not become stable, but instead has steadily drifted downward. Thus, two sources of variation are in the data. One is the process variation that we would observe if times were stable (random variation that we might observe over short periods), and the other is variation due to the downward trend in times (the large differences between times from the early 1950s compared to recent times). In using the standard deviation s of all the times between 1950 and 2002, we include both of the sources of variation. However, to determine if times in the next few years are unusual, we should consider only the process variation (variability in recent times where the process is relatively stable). Using s overestimates the process variation, resulting in control limits that are too wide to effectively signal unusually fast or slow times.

SECTION 17.3

OVERVIEW

The lower and upper specification limits (LSL and USL) for a process define the acceptable set of values of the output of a process. **Capability indexes** compare process variability σ to the process specifications. The capability index $\boldsymbol{C_p}$ is defined to be

$$C_p = \frac{\text{USL} - \text{LSL}}{6\sigma}$$

and if the process mean is μ, the capability index $\boldsymbol{C_{pk}}$ is defined to be

$$C_{pk} = \frac{|\mu - \text{nearer spec limit}|}{3\sigma}$$

We set C_{pk} to be 0 if the process mean lies outside the specification limits. Larger values of these indexes indicate higher capability. We usually do not know μ and σ so we use estimates $\hat{\mu}$ and $\hat{\sigma}$ in the formulas for C_p and C_{pk} leading to the estimates

$$\hat{C}_p = \frac{\text{USL} - \text{LSL}}{6\hat{\sigma}}$$

$$\hat{C}_{pk} = \frac{|\hat{\mu} - \text{nearer spec limit}|}{3\hat{\sigma}}$$

Interpretation of C_p and C_{pk} requires that measurements on the process output have a roughly normal distribution. These indexes are not meaningful unless the process is in control so that its center and variability are stable.

Estimates of C_p and C_{pk} can be quite inaccurate when based on small numbers of observations, due to sampling variability. You should mistrust estimates not based on at least 100 measurements.

GUIDED SOLUTIONS

Exercise 17.53

KEY CONCEPTS: Capability, C_p and C_{pk}

(a) We are given that estimates of estimate μ and σ are $\hat{\mu} = 275.065$ and $\hat{\sigma} = 38.38$. LSL = 100 and USL = 400, so we estimate C_p and C_{pk} to be

$$\hat{C}_p = \frac{\text{USL} - \text{LSL}}{6\hat{\sigma}} =$$

$$\hat{C}_{pk} = \frac{|\hat{\mu} - \text{nearer spec limit}|}{3\hat{\sigma}} =$$

(b) Specifications are now LSL = 150 and USL = 350, so now

$$\hat{C}_p = \frac{\text{USL} - \text{LSL}}{6\hat{\sigma}} =$$

$$\hat{C}_{pk} = \frac{|\hat{\mu} - \text{nearer spec limit}|}{3\hat{\sigma}} =$$

Exercise 17.59

KEY CONCEPTS: Specifications, capability indices, C_p and C_{pk}

(a) Based on the remaining 17 samples, we find

$$\bar{\bar{x}} = \text{mean of the 17 values of } \bar{x} = 43.41$$

$$s = \text{standard deviation of the remaining 85 observations} = 11.58$$

Thus, we estimate μ to be

$$\hat{\mu} = \bar{\bar{x}} = 43.41$$

and we estimate σ to be (using the fact that c_4 is very close to 1 for a sample of size 85)

$$\hat{\sigma} = s = 11.58$$

Sketch an $N(43.41, 11.38)$ distribution below, making sure to show the specification limits 54 ± 10 (or LSL = 44 and USL = 64) on the sketch.

(b) We estimate

$$\hat{C}_p = \frac{\text{USL} - \text{LSL}}{6\hat{\sigma}} =$$

$$\hat{C}_{pk} = \frac{|\hat{\mu} - \text{nearer spec limit}|}{3\hat{\sigma}} =$$

The capability is poor (both indices are very small). Why do you think this is the case? You might want to refer to the natural tolerances found in Exercise 17.33 and the sketch in part (a) of this exercise.

Exercise 17.61

KEY CONCEPTS: Percent meeting specifications, capability index C_{pk}

(a) We would estimate the process mean and standard deviation to be $\hat{\mu} = \bar{\bar{x}} = \bar{x} = 14.99$ and $\hat{\sigma} = s = 0.2239$. In its current state, therefore, measurements on the process will vary (approximately) according to an $N(14.99, 0.2239)$ distribution. The probability that a measurement x lies within the specifications 15 ± 0.5 , i.e., is between 14.5 and 15.5, is

$P(14.5 < x < 15.5)$

(b) We estimate C_{pk} to be

$$\hat{C}_{pk} = \frac{|\hat{\mu} - \text{nearer spec limit}|}{3\hat{\sigma}} =$$

COMPLETE SOLUTIONS

Exercise 17.53

(a) We are given that estimates of estimate μ and σ are $\hat{\mu} = 275.065$ and $\hat{\sigma} = 38.38$. LSL = 100 and USL = 400, so we estimate C_p and C_{pk} to be

$$\hat{C}_p = \frac{\text{USL} - \text{LSL}}{6\hat{\sigma}} = \frac{400 - 100}{6(38.38)} = 1.303$$

$$\hat{C}_{pk} = \frac{|\hat{\mu} - \text{nearer spec limit}|}{3\hat{\sigma}} = \frac{|275.065 - 400|}{3(38.38)} = 1.085$$

(b) Specifications are now LSL = 150 and USL = 350, so now

$$\hat{C}_p = \frac{\text{USL} - \text{LSL}}{6\hat{\sigma}} = \frac{350 - 150}{6(38.38)} = 0.869$$

$$\hat{C}_{pk} = \frac{|\hat{\mu} - \text{nearer spec limit}|}{3\hat{\sigma}} = \frac{|275.065 - 350|}{3(38.38)} = 0.651$$

Exercise 17.59

(a) We estimate μ to be $\hat{\mu} = \bar{\bar{x}} = 43.41$ and we estimate σ to be $\hat{\sigma} = s = 11.58$. A sketch of an $N(43.41, 11.58)$ distribution showing the specification limits 54 ± 10 (or LSL = 44 and USL = 64) is given below.

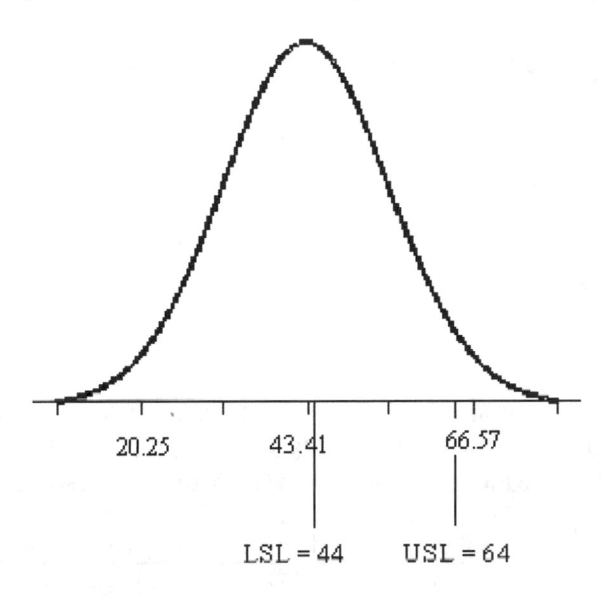

(b) We estimate

$$\hat{C}_p = \frac{\text{USL} - \text{LSL}}{6\hat{\sigma}} = \frac{64 - 44}{6(11.58)} = 0.29$$

$$\hat{C}_{pk} = \frac{|\hat{\mu} - \text{nearer spec limit}|}{3\hat{\sigma}} = \frac{|43.41 - 44|}{3(11.58)} = 0.02$$

The capability is poor (both indices are very small). The reasons are that the process is not centered (we estimate the process mean to be 43.41 but the midpoint of the specification limits is 54) and the process variability is large (we saw in Exercise 17.33 that the natural tolerances for the process are $\hat{\mu} \pm 3\hat{\sigma} = 43.41 \pm 3(11.58) = 43.41 \pm 34.74$, which is much wider than the specification limits of 54 ± 10).

Exercise 17.61

(a) We would estimate the process mean and standard deviation to be $\hat{\mu} = \bar{\bar{x}} = \bar{x} = 14.99$ and $\hat{\sigma} = s = 0.2239$. In its current state, therefore, measurements on the process will vary (approximately) according to an $N(14.99, 0.2239)$ distribution. The probability that a measurement x lies within the specifications 15 ± 0.5 , i.e., is between 14.5 and 15.5, is

$$P(14.5 < x < 15.5) = P\left(\frac{14.50 - 14.99}{0.2239} < \frac{x - 14.99}{0.2239} < \frac{15.50 - 14.99}{0.2239}\right) = P(-2.19 < Z < 2.28)$$

$$= P(Z < 2.28) - P(Z < -2.19) = 0.9887 - 0.0143 = 0.9744$$

Thus, about 97.41% of clip openings will meet specifications if the process remains in its current state.

(b) We estimate C_{pk} to be

$$\hat{C}_{pk} = \frac{|\hat{\mu} - \text{nearer spec limit}|}{3\hat{\sigma}} = \frac{|14.99 - 14.50|}{3(0.2239)} = 0.73$$

This is called six sigma quality because LSL and USL are both at least 6σ from the process mean μ.

SECTION 17.4

OVERVIEW

There are control charts for several different types of process measurements. One important type is the ***p* chart** which is a control chart based on plotting sample proportions $\hat{p}$ from regular samples from a process against the order in which the samples were taken. We estimate the process proportion p of "successes" by

$$\bar{p} = \frac{\text{total number of successes in past samples}}{\text{total number of opportunities in these samples}}$$

and then the control limits for a p chart for future samples of size n are

$$\text{UCL} = \bar{p} + 3\sqrt{\frac{\bar{p}(1-\bar{p})}{n}}, \ \text{CL} = \bar{p}, \ \text{LCL} = \bar{p} - 3\sqrt{\frac{\bar{p}(1-\bar{p})}{n}}$$

The interpretation of p charts is very similar to that of $\bar{x}$ charts. The out-of-control signals used are also the same as for $\bar{x}$ charts.

GUIDED SOLUTIONS

Exercise 17.69

KEY CONCEPTS: Center line and control limits, p chart

(a) You are told that the average number of invoices over the 10-month period is 2875 per month. Because each invoice has the potential to be unpaid, the total number of invoices in the 10-month period is the total number of opportunities for unpaid invoices. Thus,

total number of opportunities for unpaid invoices =

960 of these invoices were unpaid after 30 days (these would be our "successes" in past samples), so

$$\bar{p} = \frac{\text{total number of successes in past samples}}{\text{total number of opportunities in these samples}} =$$

(b) We expect 2875 invoices per month, so our p chart will be based on samples of size $n = 2875$. The control limits for a p chart for future samples of size $n = 2875$ are

$$\text{UCL} = \bar{p} + 3\sqrt{\frac{\bar{p}(1-\bar{p})}{n}} =$$

$$\text{CL} = \bar{p} =$$

$$\text{LCL} = \bar{p} - 3\sqrt{\frac{\bar{p}(1-\bar{p})}{n}} =$$

Exercise 17.75

KEY CONCEPTS: Center line and control limits, drawing a p chart

(a) The total number of opportunities for a student to have three or more unexcused absences from school in a month is just the total of the numbers of students each month. The total number of "successes" is just the total of the numbers of students that had three or more unexcused absences from school in a month.

$$\bar{p} = \frac{\text{total number of successes in past samples}}{\text{total number of opportunities in these samples}} =$$

We also find the average number of students per month for this 10-month period is

$\bar{n} =$

(b) Assuming $n = \bar{n} = 921.8$ each month, the control limits for a p chart for future samples of size $n = 921.8$ are

$$\text{UCL} = \bar{p} + 3\sqrt{\frac{\bar{p}(1-\bar{p})}{n}} =$$

$$\text{CL} = \bar{p} =$$

$$\text{LCL} = \bar{p} - 3\sqrt{\frac{\bar{p}(1-\bar{p})}{n}} =$$

Draw a p chart based on these control limits and the proportions with three or more unexcused absences in each of the three months. Comment on control.

(c) For October,

$$\text{UCL} = \bar{p} + 3\sqrt{\frac{\bar{p}(1-\bar{p})}{n}} =$$

$$\text{LCL} = \bar{p} - 3\sqrt{\frac{\bar{p}(1-\bar{p})}{n}} =$$

For June,

$$\text{UCL} = \bar{p} + 3\sqrt{\frac{\bar{p}(1-\bar{p})}{n}} =$$

$$\text{LCL} = \bar{p} - 3\sqrt{\frac{\bar{p}(1-\bar{p})}{n}} =$$

Add these limits to the p chart drawn in part (b). Do these exact limits affect your conclusions?

COMPLETE SOLUTIONS

Exercise 17.69

(a) The average number of invoices over the 10-month period is 2875 per month. Because each invoice has the potential to be unpaid, the total number of invoices in the 10-month period is the total number of opportunities for unpaid invoices. Thus,

$$2875 = \frac{\text{total number of opportunities for unpaid invoices}}{10}$$

so that

$$\text{total number of opportunities for unpaid invoices} = 10(2875) = 28{,}750$$

960 of these invoices were unpaid after 30 days (these would be our "successes" in past samples), so

$$\bar{p} = \frac{\text{total number of successes in past samples}}{\text{total number of opportunities in these samples}} = \frac{960}{28750} = 0.0334$$

(b) We expect 2875 invoices per month, so our p chart will be based on samples of size $n = 2875$. The control limits for a p chart for future samples of size $n = 2875$ are

$$\text{UCL} = \bar{p} + 3\sqrt{\frac{\bar{p}(1-\bar{p})}{n}} = 0.0334 + 3\sqrt{\frac{0.0334(1-0.0334)}{2875}} = 0.0334 + 0.0101 = 0.0435$$

$$\text{CL} = \bar{p} = 0.0334$$

$$\text{LCL} = \bar{p} - 3\sqrt{\frac{\bar{p}(1-\bar{p})}{n}} = 0.0334 - 3\sqrt{\frac{0.0334(1-0.0334)}{2875}} = 0.0334 - 0.0101 = 0.0233$$

Exercise 17.75

(a) The total number of opportunities for a student to have three or more unexcused absences from school in a month is just the total of the numbers of students each month. This is 911 + 947 + 939 + 942 + 918 + 920 + 931 + 925 + 902 + 883 = 9218. The total number of "successes" is just the total of the numbers of students that had three or more unexcused absences from school in a month. This is 291 + 349 + 364 + 335 + 301 + 322 + 344 + 324 + 303 + 344 = 3277. Thus,

$$\bar{p} = \frac{\text{total number of successes in past samples}}{\text{total number of opportunities in these samples}} = \frac{3277}{9218} = 0.356$$

We also find the average number of students per month for this 10-month period is

$$\bar{n} = \frac{9218}{10} = 921.8$$

(b) Assuming $n = \bar{n} = 921.8$ each month, the control limits for a p chart for future samples of size $n =$ 921.8 are

$$\text{UCL} = \bar{p} + 3\sqrt{\frac{\bar{p}(1-\bar{p})}{n}} = 0.356 + 3\sqrt{\frac{0.356(1-0.356)}{921.8}} = 0.356 + 0.0473 = 0.4033$$

$$\text{CL} = \bar{p} = 0.356$$

$$\text{LCL} = \bar{p} - 3\sqrt{\frac{\bar{p}(1-\bar{p})}{n}} = 0.356 - 3\sqrt{\frac{0.356(1-0.356)}{921.8}} = 0.356 - 0.0473 = 0.3087$$

A p chart based on these control limits is

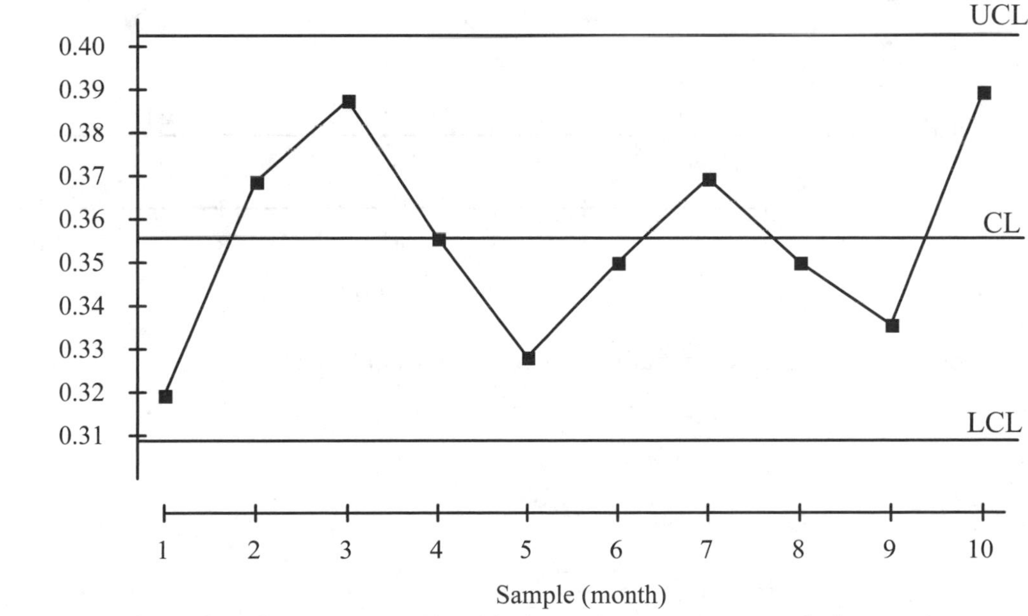

The process (proportion of students per month with three or more unexcused absences) appears to be in control, i.e., there are no months with unusually high or low proportions of absences, and no obvious trends.

(c) For October,

$$UCL = \bar{p} + 3\sqrt{\frac{\bar{p}(1-\bar{p})}{n}} = 0.356 + 3\sqrt{\frac{0.356(1-0.356)}{947}} = 0.356 + 0.0467 = 0.4027$$

$$LCL = \bar{p} - 3\sqrt{\frac{\bar{p}(1-\bar{p})}{n}} = 0.356 - 3\sqrt{\frac{0.356(1-0.356)}{947}} = 0.356 - 0.0467 = 0.3093$$

For June,

$$UCL = \bar{p} + 3\sqrt{\frac{\bar{p}(1-\bar{p})}{n}} = 0.356 + 3\sqrt{\frac{0.356(1-0.356)}{883}} = 0.356 + 0.0483 = 0.4043$$

$$LCL = \bar{p} - 3\sqrt{\frac{\bar{p}(1-\bar{p})}{n}} = 0.356 - 3\sqrt{\frac{0.356(1-0.356)}{883}} = 0.356 - 0.0483 = 0.3077$$

These limits are added to the p chart, reproduced below.

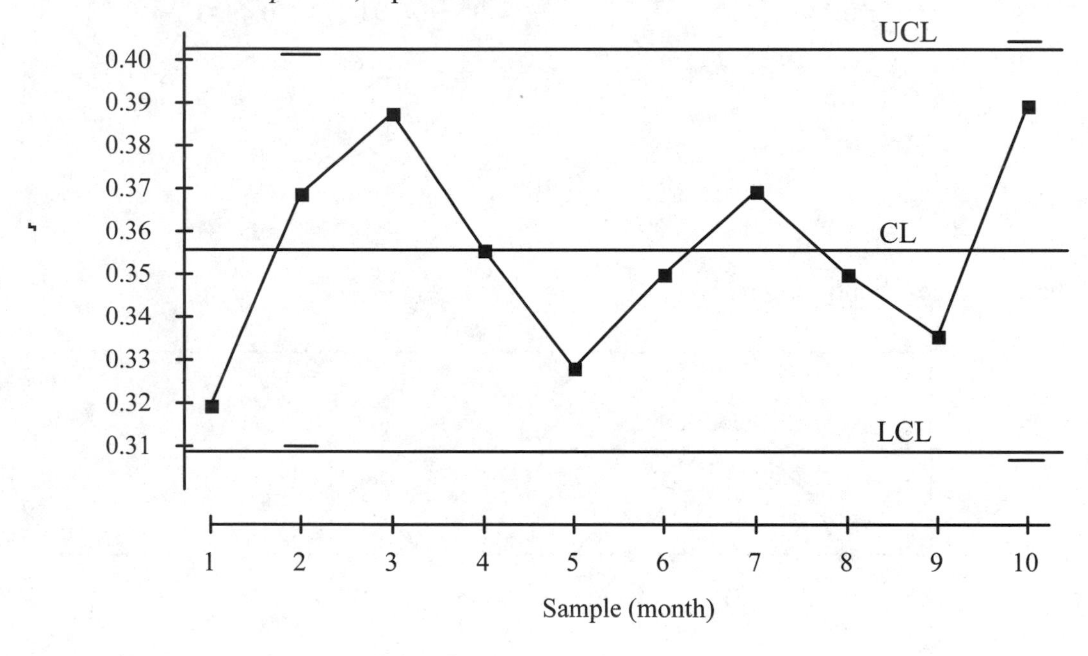

We can see that these exact limits do not affect our conclusions in this case.